High School Laboratory Manual for

Human Anatomy & Physiology

First Edition

TERRY R. MARTIN

Kishwaukee College

Mc
Graw
Hill
Education

HIGH SCHOOL LABORATORY MANUAL FOR HUMAN ANATOMY & PHYSIOLOGY: FIRST EDITION

6 7 8 9 0 QVS 1 0 9 8 7

ISBN 978–0–02–140736–1
MHID 0–02–140736–3

Vice President, Editor-in-Chief: *Marty Lange*
Vice President, EDP: *Kimberly Meriwether David*
Senior Director of Development: *Kristine Tibbetts*
Publisher: *Michael S. Hackett*
Executive Editor: *James F. Connely*
Senior Developmental Editor: *Fran Simon*
Marketing Manager: *Chris Loewenberg*
Lead Project Manager: *Peggy J. Selle*
Senior Buyer: *Sandy Ludovissy*
Senior Media Project Manager: *Tammy Juran*
Senior Designer: *Laurie B. Janssen*
Cover Designer: *Ron Bissell*
Senior Photo Research Coordinator: *John C. Leland*
Photo Research: *Danny Meldung/Photo Affairs, Inc*
Compositor: *Precision Graphics*
Typeface: *10/12 Times LT Std*
Printer: *Quad/Graphics*

www.mheonline.com

Contents

Cardiovascular System

Respiratory System

Digestive System

Urinary System

Preface

Anatomy & Physiology Lab Courses

Author Terry Martin's forty years of teaching anatomy and physiology courses, authorship of three laboratory manuals, and active involvement in the Human Anatomy and Physiology Society (HAPS) drove his determination to create a laboratory manual with an innovative approach that would benefit students. The *High School Laboratory Manual for Human Anatomy & Physiology* is written to work well with any anatomy and physiology text.

Martin Lab Manual Series . . .
Anatomy & Physiology Lab Courses

▶ Incorporates **learning outcomes and assessments** to help students master important material!

▶ **Pre-Lab** assignments are printed in the lab manual. They will help students be more prepared for lab and save instructors time during lab.

▶ **Clear, concise** writing style facilitates more thorough understanding of lab exercises.

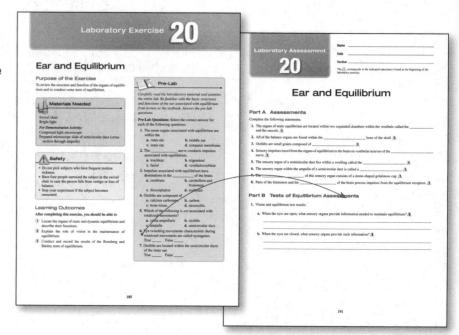

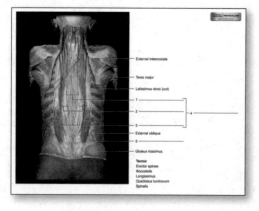

▶ Cadaver images from **Anatomy & Physiology Revealed®** (APR) are incorporated throughout the lab. Cadaver images help students make the connection from specimen to cadaver.

▶ **Micrographs** incorporated throughout the lab aid students' visual understanding of difficult topics.

▶ **Teacher's Guide** is annotated for quick and easy use by teachers and is available through your textbook's Online Learning Center. Or contact your sales representative.

Student Needs

▶ The procedures are clear, concise, and easy to follow. Relevant lists and summary tables present the contents efficiently. Histology micrographs and cadaver photos are incorporated in the appropriate locations within the associated labs.

▶ The pre-lab section now includes quiz questions. It also directs the student to carefully read the introductory material and the entire lab to become familiar with its contents. If necessary, a textbook or lecture notes might be needed to supplement the concepts.

▶ **Terminologia Anatomica** is used as the source for universal terminology in this laboratory manual. Alternative names are included when a term is introduced for the first time.

▶ Laboratory assessments immediately follow each laboratory exercise.

▶ Histology photos are placed within the appropriate laboratory exercise.

▶ A section called "Study Skills for Anatomy and Physiology" is located in the front of this laboratory manual. This section was written by students enrolled in a Human Anatomy and Physiology course.

▶ Critical Thinking Activities are incorporated within most of the laboratory exercises to enhance valuable critical thinking skills that students need throughout their lives.

▶ Cadaver images are incorporated with dissection labs.

Teacher Needs

▶ The teacher will find digital assets for use in creating customized lectures, visually enhanced tests and quizzes, and other printed support material.

▶ A correlation guide for **Anatomy & Physiology Revealed® (APR)** and the entire lab manual is available through your textbook's Online Learning Center. Cadaver images from APR are included within many of the laboratory exercises.

▶ The annotated teacher's guide for *High School Laboratory Manual for Human Anatomy and Physiology* describes the purpose of the laboratory manual and its special features, provides suggestions for presenting the laboratory exercises to students, instructional approaches, a suggested time schedule, and annotated figures and assessments. It contains a "Student Safety Contract" and a "Student Informed Consent Form." The teacher's guide is available through your textbook's Online Learning Center.

▶ Each laboratory exercise can be completed during a single laboratory session.

Educational Needs

▶ Learning outcomes with icons ⃝ have matching assessments with icons 🄰 so students can be sure they have accomplished the laboratory exercise content. Outcomes and assessments include all levels of learning skills: remember, understand, apply, analyze, evaluate, and create.

Technology

▶  Physiology Interactive Lab Simulations (Ph.I.L.S. 4.0) includes 42 lab simulations that complement and expand topics covered in the lab manual. Contact your sales representative for more information.

▶ Anatomy & Physiology Revealed (APR) is the ultimate online interactive cadaver dissection experience. This state-of-the-art program uses cadaver photos combined with a layering technique that allows the student to peel away layers of the human body to reveal structures beneath the surface. This program covers important topics from chemistry to organ systems, with animations, audio pronunciations, and comprehensive quizzing along the way. Contact your sales representative to learn more.

Guided Tour Through A Lab Exercise

The laboratory exercises include a variety of special features that are designed to stimulate interest in the subject matter, to involve students in the learning process, and to guide them through the planned activities. These features include the following:

Purpose of the Exercise
The purpose provides a statement about the intent of the exercise—that is, what will be accomplished.

Materials Needed
This section lists the laboratory materials that are required to complete the exercise and to perform the demonstrations and learning extensions.

Safety
A list of safety guidelines is included inside the front cover. Each lab session that requires special safety guidelines has a safety section. Your teacher might require some modifications of these guidelines.

Learning Outcomes
The learning outcomes list what a student should be able to do after completing the exercise. Each learning outcome will have matching assessments indicated by the corresponding icon A in the laboratory exercise or the laboratory assessment.

Laboratory Exercise 20

Ear and Equilibrium

Purpose of the Exercise
To review the structure and function of the organs of equilibrium and to conduct some tests of equilibrium.

Materials Needed
Swivel chair
Bright light
For Demonstration Activity:
Compound light microscope
Prepared microscope slide of semicircular duct (cross section through ampulla)

Safety
▶ Do not pick subjects who have frequent motion sickness.
▶ Have four people surround the subject in the swivel chair in case the person falls from vertigo or loss of balance.
▶ Stop your experiment if the subject becomes nauseated.

Learning Outcomes
After completing this exercise, you should be able to
1. Locate the organs of static and dynamic equilibrium and describe their functions.
2. Explain the role of vision in the maintenance of equilibrium.
3. Conduct and record the results of the Romberg and Bárány tests of equilibrium.

Pre-Lab
Carefully read the introductory material and examine the entire lab. Be familiar with the basic structures and functions of the ear associated with equilibrium from lecture or the textbook. Answer the pre-lab questions.

Pre-Lab Questions: Select the correct answer for each of the following questions:
1. The sense organs associated with equilibrium are within the
   a. outer ear.   b. middle ear.
   c. inner ear.   d. tympanic membrane.
2. The _____ nerve conducts impulses associated with equilibrium.
   a. trochlear   b. trigeminal
   c. facial   d. vestibulocochlear
3. Impulses associated with equilibrium have destinations in the _____ of the brain.
   a. cerebrum   b. cerebellum and brainstem
   c. diencephalon   d. midbrain
4. Otoliths are composed of
   a. calcium carbonate. b. carbon.
   c. bone tissue.   d. stereocilia.
5. Which of the following is *not* associated with rotational movements?
   a. crista ampullaris   b. otoliths
   c. ampulla   d. semicircular duct
6. Eye twitching movements characteristic during rotational movements are called nystagmus.
   True _____ False _____
7. Otoliths are located within the semicircular ducts of the inner ear.
   True _____ False _____

185

Pre-Lab
The pre-lab includes quiz questions and directs the student to carefully read introductory material and examine the entire laboratory contents after becoming familiar with the topics from a textbook or lecture. After successfully answering the pre-lab questions, the student is prepared to become involved in the laboratory exercise.

Introduction
The introduction describes the subject of the exercise or the ideas that will be investigated. It includes all of the information needed to perform the laboratory exercise.

Procedure
The procedure provides a set of detailed instructions for accomplishing the planned laboratory activities. Usually these instructions are presented in outline form so that a student can proceed efficiently through the exercise in stepwise fashion.

The procedures include a wide variety of laboratory activities and, from time to time, direct the student to complete various tasks in the laboratory assessments.

The sense of equilibrium involves two sets of sensory organs. One set helps to maintain the stability of the head and body when they are motionless or during linear acceleration and produces a sense of static (gravitational) equilibrium. The other set is concerned with balancing the head and body when angular acceleration produces a sense of dynamic (rotational) equilibrium.

The sense organs associated with the sense of static equilibrium are located within the vestibules of the inner ears. Two chambers, the utricle and saccule, contain receptors called maculae. Each macula is composed of hair cells embedded within a gelatinous otolithic membrane. Embedded within the otolithic membrane are numerous tiny calcium carbonate "ear stones" called otoliths. As we change our head position, gravitational forces allow the shift of the otoliths to bend the stereocilia of the hair cells of the macula, resulting in stimulation of sensory vestibular neurons. Stimulation of the maculae also occurs during linear acceleration. Horizontal acceleration (as occurs when riding in a car) involves the utricle; vertical acceleration (as occurs when riding in an elevator) involves the saccule. As a result of static equilibrium, recognition of movements such as falling are detected and postural adjustments are accomplished.

The sense organs associated with the sense of dynamic equilibrium are located within the ampullae of the three semicircular ducts of the inner ear. Each membranous semicircular duct is located within a semicircular canal of the temporal bone. A small elevation within each ampulla possesses the crista ampullaris. Each crista ampullaris contains hair cells with stereocilia embedded within a gelatinous cap called the cupula. During rotational movements, the cupula is bent, which stimulates hair cells and sensory neurons of the vestibular nerve. Because the three semicircular ducts are in different planes, rotational movements in any direction result in stimulation of the associated hair cells. Impulses from vestibular neurons of the semicircular ducts result in reflex movements of the eye. During rotational movements, characteristic twitching movements of the eyes called nystagmus occur, often accompanied by dizziness (vertigo).

Impulses from inner ear receptors travel over the vestibular neurons of the vestibulocochlear nerve and include

Demonstration Activity
Observe the cross section of the semicircular duct through the ampulla in the demonstration microscope. Note the crista ampullaris (fig. 20.5) projecting into the lumen of the membranous labyrinth, which in a living person is filled with endolymph. The space between the membranous and bony labyrinths is filled with perilymph.

Procedure B—Tests of Equilibrium
Use figures 20.1 and 20.3 as references as you progress through the various tests of equilibrium. Perform the following tests, using a person as a test subject who is not easily disturbed by dizziness or rotational movement. Also have some other students standing close by to help prevent the test subject from falling during the tests. *The tests should be stopped immediately if the test subject begins to feel uncomfortable or nauseated.*

1. *Vision and equilibrium test.* To demonstrate the importance of vision in the maintenance of equilibrium, follow these steps:
   a. Have the test subject stand erect on one foot for 1 minute with his or her eyes open.
   b. Observe the subject's degree of unsteadiness.
   c. Repeat the procedure with the subject's eyes closed. *Be prepared to prevent the subject from falling.*
   d. Answer the questions related to the vision and equilibrium test in Part B of the laboratory assessment.
2. *Romberg test.* The purpose of this test is to evaluate how the organs of static equilibrium in the vestibule enable one to maintain balance (fig. 20.1). To conduct this test, follow these steps:
   a. Position the test subject close to a chalkboard with the back toward the board.
   b. Place a bright light in front of the subject so that a shadow of the body is cast on the board.
   c. Have the subject stand erect with feet close together and eyes staring straight ahead for 3 minutes.

Demonstration Activities Demonstration activities appear in separate boxes. They describe specimens, specialized laboratory equipment, or other materials of interest that a teacher may want to display to enrich the student's laboratory experience.

Demonstration Activity

Examine a fresh chicken bone and a chicken bone that has been soaked for several days in vinegar or overnight in dilute hydrochloric acid. Wear disposable gloves for handling these bones. This acid treatment removes the inorganic salts from the bone extracellular matrix. Rinse the bones in water and note the texture and flexibility of each (fig. 12.7a). The bone becomes soft and flexible without the support of the inorganic salts with calcium.

Examine the specimen of chicken bone that has been exposed to high temperature (baked at 121°C/250°F for 2 hours). This treatment removes the ... from the bone ... bone comes ... the collagen ... of the quali- ... provides ten- ... the chicken

Learning Extension Activities Learning extension activities also appear in separate boxes. They encourage students to extend their laboratory experiences. Some of these activities are open-ended in that they suggest the student plan an investigation or experiment and carry it out after receiving approval from the laboratory teacher. Some of the figures are illustrated as line art or in grayscale. This will allow colored pencils to be used as a visual learning activity to distinguish various structures.

Learning Extension Activity

Repeat the demonstration of diffusion using a petri dish filled with ice-cold water and a second dish filled with very hot water. At the same moment, add a crystal of potassium permanganate to each dish and observe the circle as before. What difference do you note in the rate of diffusion in the two dishes? How do you explain this difference? _____

_____

_____

Illustrations Diagrams similar to those in a textbook often are used as aids for reviewing subject matter. Other illustrations provide visual instructions for performing steps in procedures or are used to identify parts of instruments or specimens. Micrographs are included to help students identify microscopic structures or to evaluate student understanding of tissues.

In some exercises, the figures include line drawings suitable for students to color with colored pencils. This activity may motivate students to observe the illustrations more carefully and help them to locate the special features represented in the figures.

Hyoid bone

Laboratory Assessments A laboratory assessment form to be completed by the student immediately follows each exercise. These assessments include various types of review activities, spaces for sketches of microscopic objects, tables for recording observations and experimental results, and questions dealing with the analysis of such data.

As a result of these activities, students will develop a better understanding of the structural and functional characteristics of their bodies and will increase their skills in gathering information by observation and experimentation. By completing all of the assessments, students will be able to determine if they were able to accomplish all of the learning outcomes.

Laboratory Assessment

20

Name _____
Date _____
Section _____
The △ corresponds to the indicated outcome(s) found at the beginning of the laboratory exercise.

Ear and Equilibrium

Part A Assessments

Complete the following statements:

1. The organs of static equilibrium are located within two expanded chambers within the vestibule called the _____ and the saccule. △
2. All of the balance organs are found within the _____ bone of the skull. △
3. Otoliths are small grains composed of _____. △
4. Sensory impulses travel from the organs of equilibrium to the brain on vestibular neurons of the _____ nerve. △
5. The sensory organ of a semicircular duct lies within a swelling called the _____. △
6. The sensory organ within the ampulla of a semicircular duct is called a _____. △
7. The _____ of this sensory organ consists of a dome-shaped gelatinous cap. △
8. Parts of the brainstem and the _____ of the brain process impulses from the equilibrium receptors. △

Part B Tests of Equilibrium Assessments

1. Vision and equilibrium test results:

... open, what sensory organs provide information needed to maintain equilibrium? △

_____

... closed, what sensory organs provide such information? △

_____

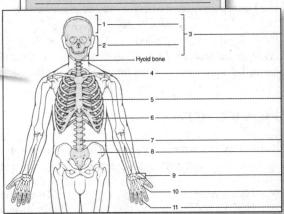

Ampulla
Cupula
Hair
Hair cell
Region of sensory nerve fibers

Histology Histology photos placed within the appropriate exercise.

Acknowledgments

I value all the support and encouragement by the staff at McGraw-Hill, including Michelle Watnick, Jim Connely, Fran Simon, and Peggy Selle. Special recognition is granted to Colin Wheatley for his insight, confidence, wisdom, warmth, and friendship

I am grateful for the professional talent of Lurana Bain, Phillip Snider, and Janet Brodsky for their significant involvement in the development of some of the laboratory exercises. The main contents of the surface anatomy lab were provided by Lurana Bain, Ph.I.L.S. lab simulations components by Phillip Snider, and BIOPAC® labs by Janet Brodsky. I also want to thank Phillip Snider, Sue Caley-Opsal, and Kash Dutta for their contributions to the digital content included with this lab manual.

I am particularly thankful to Dr. Norman Jenkins and Dr. David Louis, retired presidents of Kishwaukee College, and Dr. Thomas Choice, president of Kishwaukee College, for their support, suggestions, and confidence in my endeavors. I am appreciative for the expertise of Womack Photography for numerous contributions. The professional reviews of the nursing procedures were provided by Kathy Schnier. I am also grateful to Laura Anderson, Rebecca Doty, Michele Dukes, Troy Hanke, Jenifer Holtzclaw, Stephen House, Shannon Johnson, Brian Jones, Marissa Kannheiser, Morgan Keen, Marcie Martin, Angele Myska, Sparkle Neal, Bonnie Overton, Susan Rieger, Eric Serna, Robert Stockley, Shatina Thompson, Nancy Valdivia, Marla Van Vickle, Jana Voorhis, Joyce Woo, and DeKalb Clinic for their contributions. There have been valuable contributions from my students, who have supplied thoughtful suggestions and assisted in clarification of details.

To my son Ross, an art instructor, I owe gratitude for his keen eye and creative suggestions. Foremost, I am appreciative to Sherrie Martin, my spouse and best friend, for advice, understanding, and devotion throughout the writing and revising.

Terry R. Martin
Kishwaukee College
21193 Malta Road
Malta, IL 60150

Reviewers

I would like to express my sincere gratitude to all reviewers of the laboratory manual who provided suggestions for its improvement. Their thoughtful comments and valuable suggestions are greatly appreciated. They include the following:

M. Abdel-Sayed
MEC/CUNY
Sharon I. Allen
Reading Area Community College
Frank Ambriz
University of Texas–Pan American
Pamela Anderson Cole
Shelton State Community College
Bert Atsma
Union County College
Jerry D. Barton II
Tarrant County College–South Campus
Rachel Beecham
Mississippi Valley State University
Moges Bizuneh
Ivy Tech Community College
Danita Bradshaw-Ward
Eastfield College
Gary Brady
Spokane Falls Community College
Janet Brodsky
Ivy Tech Community College
Beth Campbell
Itawamba Community College
Ronald Canterbury
University of Cincinnati

Claire Carpenter
Yakima Valley Community College
Roger Choate
Oklahoma City Community College
Ana Christensen
Lamar University
Larissa Clark
Arkansas State University-Newport
Harry Davis
Gloucester County College
Mary Beth Davison
Chatham University
Frank J. DeMaria
Suffolk County Community College
William E. Dunscombe
Union County College
Scott Elliott
Whatcom Community College
Marirose T. Ethington
Genesee Community College
David L. Evans
Pennsylvania College of Technology
Amy Fenech Sandy
Columbus Technical College
Tammy Filliater
MSU–Great Falls College of Technology

Vanessa A. Fitsanakis
King College
Andrew Flick
Blue Ridge Community College
Katelijne Flies
Central New Mexico Community College
Maria Florez
Lone Star College–Cy Fair
Mary C. Fox
University of Cincinnati
Sharon Fugate
Madisonville Community College
Tammy R. Gamza
Arkansas State University Beebe
Joseph D. Gar
West Kentucky Community and Technical College
Kristine Garner
University of Arkansas–Fort Smith
Anthony J. Gaudin
Ivy Tech Community College
Samita Ghoshal
Lone Star College
Louis A. Giacinti
Milwaukee Area Technical College

Tejendra Gill
University of Houston

Gary Glaser
Erie Community College

Brent M. Graves
Northern Michigan University

Dale Harrington
Caldwell Community College

Gillian Hart
Trinidad State Junior College

Clare Hays
Metro State College of Denver

Gerald A. Heins
Northeast Wisconsin Technical College

Todd Heldreth
Charleston Southern University

D.J. Hennager
Kirkwood Community College

Kerrie Hoar
University of Wisconsin–La Crosse

Dale R. Horeth
Tidewater Community College

Michele Iannuzzi Sucich
SUNY Orange

Mark Jaffe
Nova Southeastern University

Dena Johnson
Tarrant County College NW

Edward Johnson
Central Oregon Community College

Jody Johnson
Arapahoe Comm. College

Geeta Joshi
Southeastern Community College

Susanne Kalup
Westmoreland County Community College

Lisa A. E. Kaplan
Quinnipiac University

Robert Keeton
University of Arkansas Community College

Suzanne Kempke
St. John's River Community College

L. Henry Kermott
St. Olaf College

Steven Kish
Zane State College

Will Kleinelp
Middlesex County College

Leigh Kleinert
Grand Rapids Community College

Michael S. Kopenitis
Amarillo College

Gina Langley
ENMU-Ruidoso

J. Ellen Lathrop-Davis
CCBC-Catonsville

Steven Leadon
Durham Technical Community College

Leigh Levitt
Union County College

Jerri K. Lindsey
Tarrant County College-Northeast Campus

Lynn B. Littman
Gwynedd-Mercy College
Community College of Philadelphia

Kenneth Long
California Lutheran University

Sarah Lovern
Concordia University

Jaime Malcore Tjossem
Rochester Community and Technical College

Peter Malo
Richard J Daley College

Anita Mandal
Edward Waters College

Barry Markillie
Cape Fear Community College

Deborah McCool
Mount Aloysius College

Lori McGrew
Belmont

Cherie McKeever
MSU College of Technology

Jonathan McMemamin-Balano
Norwalk Community College

Mark E. Meade
Jacksonville State University

Judy Megaw
Indian River State College

Claire Miller
Community College of Denver

Howard K. Motoike
LaGuardia Community College

Necia Nicholas
Calhoun Community College

Nancy O'Keefe
Purdue University Calumet

Sidney Palmer
Brigham Young University–Idaho

Karen Payne
Chattanooga State Technical Community College

Roger D. Peffer
Montana State University–Great Falls College of Technology

Andrew J. Petto
University of Wisconsin–Milwaukee

Danny M. Pincivero
The University of Toledo

Sara Reed Houser
Jefferson College of Health Sciences

Jackie Reynolds
Richland College

Rebecca Roush
Sandhills Community College

Tim Roye
San Jacinto College

Charlotte Russell
Simmons College

Radmila Sarac
Purdue University Calumet

Kathleen Sellers
Columbus State University

Kelly J. Sexton
Dallas County Community College District

Donald Shaw
University of Tennessee at Martin

Mark A. Shoop
Tennessee Wesleyan College

Crystal Sims
Cossatot Community College UA

Michelle Slover
Clarke College

Phillip Snider
Gadsden State Community College

Olufemi Sodeinde
Medgar Evers College, CUNY

Asha Stephens
College of the Mainland

Lydia Thebeau
Missouri Baptist University

Janis Thompson
Lorain County Community College

Diane G. Tice
Morrisville State College

Terri L. Tillen
Madisonville Community College

Pete Van Dyke
Walla Walla Community College

Delon Washo-Krupps
Arizona State University

Clement Yedjou
Jackson State University

Ruth A. Young
Northern Essex Community College

Paul Zillgitt
University of Wisconsin-Waukesha

David J. Zimmer
Erie Community College

Jay Zimmer
South Florida Community College

Jeffrey Zuiderveen
Columbus State University

About the Author

This laboratory manual series is by Terry R. Martin of Kishwaukee College. Terry's teaching experience of over forty years, his interest in students and love for college instruction, and his innovative attitude and use of technology-based learning enhance the solid tradition of his other well-established laboratory manuals. Among Terry's awards are the Kishwaukee College Outstanding Educator, Phi Theta Kappa Outstanding Instructor Award, Kishwaukee College ICCTA Outstanding Educator Award, Who's Who Among America's Teachers, Kishwaukee College Faculty Board of Trustees Award of Excellence, and Continued Excellence Award for Phi Theta Kappa Advisors. Terry's professional memberships include the National Association of Biology Teachers, Illinois Association of Community College Biologists, Human Anatomy and Physiology Society, former Chicago Area Anatomy and Physiology Society (founding member), Phi Theta Kappa (honorary member), and Nature Conservancy. In addition to writing many publications, he co-produced with Hassan Rastegar a videotape entitled *Introduction to the Human Cadaver and Prosection,* published by Wm. C. Brown Publishers. Terry revised the *Laboratory Manual to Accompany Hole's Human Anatomy and Physiology,* Thirteenth Edition, revised the *Laboratory Manual to Accompany Hole's Essentials of Human Anatomy and Physiology,* Eleventh Edition, and authored *Human Anatomy and Physiology Laboratory Manual, Fetal Pig Dissection*, Third Edition. A series of seven LabCam videos of anatomy and physiology laboratory processes were produced at Kishwaukee College and published by McGraw-Hill Higher Education. Cadaver dissection experiences have been provided for

his students for over twenty-five years. Terry teaches portions of EMT and paramedic classes and serves as a Faculty Consultant for Advanced Placement Biology examination readings. Terry has also been a faculty exchange member in Ireland. The author locally supports historical preservation, natural areas, scouting, and scholarship. We are pleased to have Terry continue the tradition of authoring laboratory manuals for McGraw-Hill Higher Education and McGraw-Hill School Education.

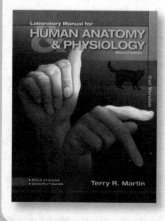
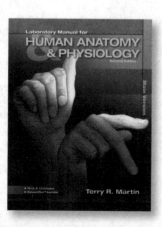

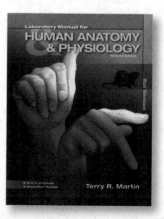

To the Student

The exercises in this laboratory manual will provide you with opportunities to observe various anatomical structures and to investigate certain physiological phenomena. Such experiences should help you relate specimens, models, microscope slides, and your body to what you have learned in the lecture and read about in the textbook.

Frequent variations exist in anatomical structures among humans. The illustrations in the laboratory manual represent normal (normal means the most common variation) anatomy. Variations from normal anatomy do not represent abnormal anatomy unless some function is impaired.

The following list of suggestions and study skills may make your laboratory activities more effective and profitable.

1. Prepare yourself before attending the laboratory session by reading the assigned exercise and reviewing the related sections of the textbook and lecture notes as indicated in the pre-lab section of the laboratory exercise. Answer the pre-lab questions. It is important to have some understanding of what will be done in the lab before you come to class.

2. Be on time. During the first few minutes of the laboratory meeting, the teacher often will provide verbal instructions. Make special note of any changes in materials to be used or procedures to be followed. Also listen carefully for information about special techniques to be used and precautions to be taken.

3. Keep your work area clean and your materials neatly arranged so that you can locate needed items. This will enable you to efficiently proceed and will reduce the chances of making mistakes.

4. Pay particular attention to the purpose of the exercise, which states what you are to accomplish in general terms, and to the learning outcomes, which list what you should be able to do as a result of the laboratory experience. Then, before you leave the class, review the outcomes and make sure that you can perform all of the assessments.

5. Precisely follow the directions in the procedure and proceed only when you understand them clearly. Do not improvise procedures unless you have the approval of the laboratory teacher. Ask questions if you do not understand exactly what you are supposed to do and why you are doing it.

6. Handle all laboratory materials with care. These materials often are fragile and expensive to replace. Whenever you have questions about the proper treatment of equipment, ask the teacher.

7. Treat all living specimens humanely and try to minimize any discomfort they might experience.

8. Although at times you might work with a laboratory partner or a small group, try to remain independent when you are making observations, drawing conclusions, and completing the activities in the laboratory reports.

9. Record your observations immediately after making them. In most cases, such data can be entered in spaces provided in the laboratory assessments.

10. Read the instructions for each section of the laboratory assessment before you begin to complete it. Think about the questions before you answer them. Your responses should be based on logical reasoning and phrased in clear and concise language.

11. At the end of each laboratory period, clean your work area and the instruments you have used. Return all materials to their proper places and dispose of wastes, including glassware or microscope slides that have become contaminated with human blood or body fluids, as directed by the laboratory instructor. Wash your hands thoroughly before leaving the laboratory.

Study Skills for Anatomy and Physiology

Students have found that certain study skills worked well for them while enrolled in Anatomy and Physiology. Although everyone has his or her learning style, there are techniques that work well for most students. Using some of the skills listed here could make your course more enjoyable and rewarding.

1. **Time management:** Prepare monthly, weekly, and daily schedules. Include dates of quizzes, exams, and projects on the calendar. On your daily schedule, budget several short study periods. Daily repetition alleviates cramming for exams. Prioritize your time so that you still have time for work and leisure activities. Find an appropriate study atmosphere with minimum distractions.

2. **Note taking:** Look for the main ideas and briefly express them in your own words. Organize, edit, and review your notes soon after the lecture. Add textbook information to your notes as you reorganize them. Underline or highlight with different colors the important points,

To the Student

major headings, and key terms. Study your notes daily, as they provide sequential building blocks of the course content.

3. **Chunking:** Organize information into logical groups or categories. Study and master one chunk of information at a time. For example, study the bones of the upper limb, lower limb, trunk, and head as separate study tasks.

4. **Mnemonic devices:** An *acrostic* is a combination of association and imagery to aid your memory. It is often in the form of a poem, rhyme, or jingle in which the first letter of each word corresponds to the first letters of the words you need to remember. **So Long Top Part, Here Comes The Thumb** is an example of such a mnemonic device for remembering the eight carpals in a correct sequence. *Acronyms* are words formed by the first letters of the items to remember. *IPMAT* is an example of this type of mnemonic device to help you remember the phases of the cell cycle in the correct sequence. Try to create some of your own.

5. **Note cards/flash cards:** Make your own. Add labels and colors to enhance the material. Keep them with you in your pocket or purse. Study them often and for short periods. Concentrate on a small number of cards at one time. Shuffle your cards and have someone quiz you on their content. As you become familiar with the material, you can set aside cards that don't require additional mastery.

6. **Recording and recitation:** An auditory learner can benefit by recording lectures and review sessions with a cassette recorder. Many students listen to the taped sessions as they drive or just before going to bed. Reading your notes aloud can help also. Explain the material to anyone (even if there are no listeners). Talk about anatomy and physiology in everyday conversations.

7. **Study groups:** Small study groups that meet periodically to review course material and compare notes have helped and encouraged many students. However, keep the group on the task at hand. Work as a team and alternate leaders. This group often becomes a support group.

Practice sound study skills during your anatomy and physiology endeavor.

The Use of Animals in Biology Education*

The National Association of Biology Teachers (NABT) believes that the study of organisms, including nonhuman animals, is essential to the understanding of life on Earth. NABT recommends the prudent and responsible use of animals in the life science classroom. NABT believes that biology teachers should foster a respect for life. Biology teachers also should teach about the interrelationship and interdependency of all things.

Classroom experiences that involve nonhuman animals range from observation to dissection. NABT supports these experiences so long as they are conducted within the long-established guidelines of proper care and use of animals, as developed by the scientific and educational community.

As with any instructional activity, the use of nonhuman animals in the biology classroom must have sound educational objectives. Any use of animals, whether for observation or dissection, must convey substantive knowledge of biology. NABT believes that biology teachers are in the best position to make this determination for their students.

NABT acknowledges that no alternative can substitute for the actual experience of dissection or other use of animals and urges teachers to be aware of the limitations of alternatives. When the teacher determines that the most effective means to meet the objectives of the class do not require dissection, NABT accepts the use of alternatives to dissection, including models and the various forms of multimedia. The Association encourages teachers to be sensitive to substantive student objections to dissection and to consider providing appropriate lessons for those students where necessary.

To implement this policy, NABT endorses and adopts the "Principles and Guidelines for the Use of Animals in Precollege Education" of the Institute of Laboratory Animals Resources (National Research Council). Copies of the "Principles and Guidelines" may be obtained from the ILAR (2101 Constitution Avenue, NW, Washington, DC 20418; 202-334-2590).

*Adopted by the Board of Directors in October 1995. This policy supersedes and replaces all previous NABT statements regarding animals in biology education.

Scientific Method and Measurements

Purpose of the Exercise

To become familiar with the scientific method of investigation, to learn how to formulate sound conclusions, and to provide opportunities to use the metric system of measurements.

Materials Needed

Meterstick
Calculator
Human skeleton

Learning Outcomes

After completing this exercise, you should be able to

1. Convert English measurements to the metric system, and vice versa.

2. Measure and record upper limb lengths and heights of ten subjects.

3. Apply the scientific method to test the validity of a hypothesis concerning the direct, linear relationship between human upper limb length and height.

4. Design an experiment, formulate a hypothesis, and test it using the scientific method.

Pre-Lab

Carefully read the introductory material and examine the entire lab. Be familiar with the scientific method from lecture or the textbook. Answer the pre-lab questions.

Pre-Lab Questions: Select the correct answer for each of the following questions:

1. To explain scientific biological phenomena, scientists use a technique called
   a. the scientific method.   b. the scientific law.
   c. conclusions.   d. measurements.

2. Which of the following represents the correct sequence of the scientific method?
   a. analysis of data, conclusions, observations, experiment, hypothesis
   b. conclusions, experiment, hypothesis, analysis of data, observations
   c. observations, hypothesis, experiment, analysis of data, conclusions
   d. hypothesis, observations, experiment, analysis of data, conclusions

3. A theory, verified continuously from experiments, might become known as the
   a. conclusions.   b. hypothesis.
   c. valid results.   d. scientific law.

4. The most likely scientific unit for measuring the height of a person would be
   a. feet.   b. centimeters.
   c. inches.   d. kilometers.

5. Which of the following is *not* a unit of the metric system of measurements?
   a. centimeters   b. liters
   c. inches   d. millimeters

6. The hypothesis is formulated from the results of the experiment.
   True _____ False _____

7. A centimeter represents an example of a metric unit of length.
   True _____ False _____

Scientific investigation involves a series of logical steps to arrive at explanations for various biological phenomena. This technique, called the *scientific method,* is used in all disciplines of science. It allows scientists to draw logical and reliable conclusions about phenomena.

The scientific method begins with *observations* related to the topic under investigation. This step commonly involves the accumulation of previously acquired information and/or your observations of the phenomenon. These observations are used to formulate a tentative explanation known as the *hypothesis.* An important attribute of a hypothesis is that it must be testable. The testing of the hypothesis involves performing a carefully controlled *experiment* to obtain data that can be used to support, reject, or modify the hypothesis. An *analysis of data* is conducted using sufficient information collected during the experiment. Data analysis may include organization and presentation of data as tables, graphs, and drawings. From the interpretation of the data analysis, *conclusions* are drawn. (If the data do not support the hypothesis, you must reexamine the experimental design and the data, and if needed develop a new hypothesis.) The final presentation of the information is made from the conclusions. Results and conclusions are presented to the scientific community for evaluation through peer reviews, presentations at professional meetings, and published articles. If many investigators working independently can validate the hypothesis by arriving at the same conclusions, the explanation becomes a **theory.** A theory verified continuously over time and accepted by the scientific community becomes known as a **scientific law** or **principle.** A scientific law serves as the standard explanation for an observation unless it is disproved by new information. The five components of the scientific method are summarized as

Observations

↓

Hypothesis

↓

Experiment

↓

Analysis of data

↓

Conclusions

Metric measurements are characteristic tools of scientific investigations. The English system of measurements is often used in the United States, so the investigator must make conversions from the English system to the metric system. Use table 1.1 for the conversion of English units of measure to metric units for length, mass, volume, time, and temperature.

Procedure A—Using the Steps of the Scientific Method

To familiarize you with the components of the scientific method, this procedure represents a specific example of the order of the steps utilized. Each of the steps for this procedure will guide you through the proper sequence in an efficient pathway.

1. Many people have observed a correlation between the length of the upper and lower limbs and the height (stature) of an individual. For example, a person who has long upper limbs (the arm, forearm, and hand combined) tends to be tall. Make some visual observations of other people in your class to observe a possible correlation.
2. From such observations, the following hypothesis is formulated: The length of a person's upper limb is equal to 0.4 (40%) of the height of the person. Test this hypothesis by performing the following experiment.
3. In this experiment, use a meterstick (fig. 1.1) to measure an upper limb length of ten subjects. For each measurement, place the meterstick in the axilla (armpit) and record the length in centimeters to the end of the longest finger (fig. 1.2). Obtain the height of each person in centimeters by measuring them without shoes against a wall (fig. 1.3). The height of each person can be calculated by multiplying each individual's height in inches by 2.54 to obtain his/her height in centimeters. Record all your measurements in Part A of Laboratory Assessment 1.
4. The data collected from all of the measurements can now be analyzed. The expected (predicted) correlation between upper limb length and height is determined using the following equation:

$$\text{Height} \times 0.4 = \text{expected upper limb length}$$

FIGURE 1.1 Metric ruler with metric lengths indicated. A meterstick length would be 100 centimeters. (The image size is approximately to scale.)

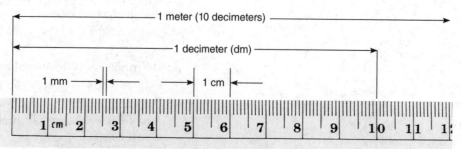

Metric ruler

TABLE 1.1 Metric Measurement System and Conversions

Measurement	Unit & Abbreviation	Metric Equivalent	Conversion Factor Metric to English (approximate)	Conversion Factor English to Metric (approximate)
Length	1 kilometer (km)	1,000 (10^3) m	1 km = 0.62 mile	1 mile = 1.61 km
	1 meter (m)	100 (10^2) cm 1,000 mm	1 m = 1.1 yards = 3.3 feet = 39.4 inches	1 yard = 0.9 m 1 foot = 0.3 m
	1 decimeter (dm)	0.1 (10^{-1}) m	1 dm = 3.94 inches	1 inch = 0.25 dm
	1 centimeter (cm)	0.01 (10^{-2}) m	1 cm = 0.4 inches	1 foot = 30.5 cm 1 inch = 2.54 cm
	1 millimeter (mm)	0.001 (10^{-3}) m 0.1 cm	1 mm = 0.04 inches	
	1 micrometer (μm)	0.000001 (10^{-6}) m 0.001 mm		
Mass	1 metric ton (t)	1,000 kg	1 t = 1.1 ton	1 ton = 0.91 t
	1 kilogram (kg)	1,000 g	1 kg = 2.2 pounds	1 pound = 0.45 kg
	1 gram (g)	1,000 mg	1 g = 0.04 ounce	1 pound = 454 g 1 ounce = 28.35 g
	1 milligram (mg)	0.001 g		
Volume (liquids and gases)	1 liter (L)	1,000 mL	1 L = 1.06 quarts	1 gallon = 3.78 L 1 quart = 0.95 L
	1 milliliter (mL)	0.001 L 1 cubic centimeter (cc or cm^3)	1 mL = 0.03 fluid ounce 1 mL = 1⁄4 teaspoon 1 mL = 15–16 drops	1 quart = 946 mL 1 fluid ounce = 29.6 mL 1 teaspoon = 5 mL
Time	1 second (s)	1⁄60 minute	same	same
	1 millisecond (ms)	0.001 s	same	same
Temperature	Degrees Celsius (°C)		°F = 9/5 °C + 32	°C = 5/9 (°F − 32)

FIGURE 1.2 Measurement of upper limb length.

The observed (actual) correlation to be used to test the hypothesis is determined by

Length of upper limb/height = actual % of height

5. A graph is an excellent way to display a visual representation of the data. Plot the subjects' data in Part A of the laboratory assessment. Plot the upper limb length of each subject on the x-axis and the height of each person on the y-axis. A line is already located on the graph that represents a hypothetical relationship of 0.4 (40%) upper limb length compared to height. This is a graphic representation of the original hypothesis.

6. Compare the distribution of all of the points (actual height and upper limb length) that you placed on the graph with the distribution of the expected correlation represented by the hypothesis.

7. Complete Part A of the laboratory assessment.

Procedure B—Design an Experiment

You have completed the steps of the scientific method with guidance directions in Procedure A. This procedure will allow for less guidance and more flexibility using the scientific method.

FIGURE 1.3 Measurement of height.

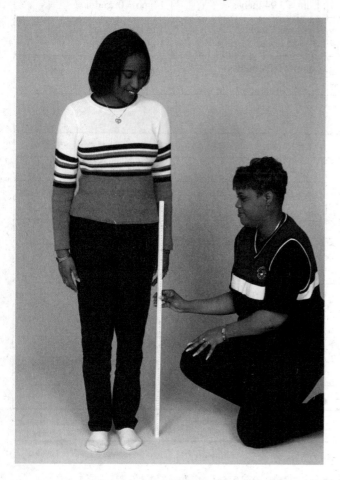

Critical Thinking Activity

You have probably concluded that there is some correlation of the length of body parts to height. Often when a skeleton is found, it is not complete, especially when paleontologists discover a skeleton. It is occasionally feasible to use the length of a single bone to estimate the height of an individual. Observe human skeletons and locate the radius bone in the forearm. Use your observations to identify a mathematical relationship between the length of the radius and height. Formulate a hypothesis that can be tested. Make measurements, analyze data, and develop a conclusion from your experiment. Complete Part B of the laboratory assessment.

Laboratory Assessment 1

Name ______________________

Date ______________________

Section ______________________

The A corresponds to the indicated outcome(s) found at the beginning of the laboratory exercise.

Scientific Method and Measurements

Part A Assessments

1. Record measurements for the upper limb length and height of ten subjects. Use a calculator to determine the expected upper limb length and the actual percentage (as a decimal or a percentage) of the height for the ten subjects. Record your results in the following table. 2

Subject	Measured Upper Limb Length (cm)	Height* (cm)	Height × 0.4 = Expected Upper Limb Length (cm)	Actual % of Height = Measured Upper Limb Length (cm)/Height (cm)
1.				
2.				
3.				
4.				
5.				
6.				
7.				
8.				
9.				
10.				

*The height of each person can be calculated by multiplying each individual's height in inches by 2.54 to obtain his/her height in centimeters. 1

2. Plot the distribution of data (upper limb length and height) collected for the ten subjects on the following graph. The line located on the graph represents the *expected* 0.4 (40%) ratio of upper limb length to measured height (the original hypothesis). (The x-axis represents upper limb length, and the y-axis represents height.) Draw a line of *best fit* through the distribution of points of the plotted data of the ten subjects. Compare the two distributions (expected line and the distribution line drawn for the ten subjects). **3**

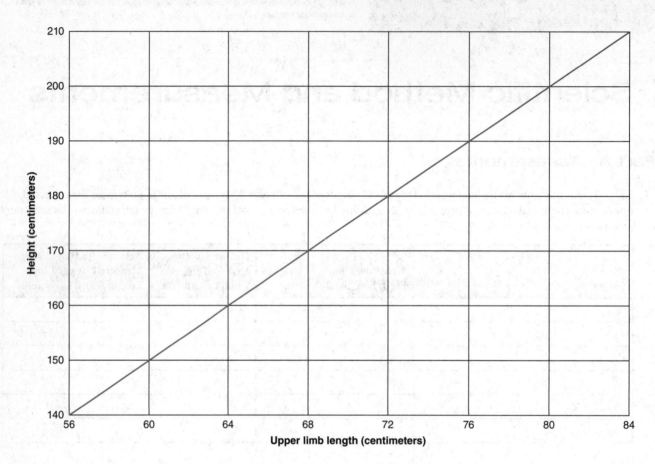

3. Does the distribution of the ten subjects' measured upper limb lengths support or reject the original hypothesis? _____ Explain your answer. **3**

Part B Assessments

1. Describe your observations of a possible correlation between the radius length and height. /A\

2. Write a hypothesis based on your observations. /A\

3. Describe the design of the experiment that you devised to test your hypothesis. /A\

4. Place your analysis of the data in this space in the form of a table and a graph. /A\

   a. Table:

   b. Graph:

5. Based on an analysis of your data, what conclusions can you make? Did these conclusions confirm or refute your original hypothesis? 𝐀

6. Discuss your results and conclusions with classmates. What common conclusion can the class formulate about the correlation between radius length and height? 𝐀

2

Cell Structure and Function

Purpose of the Exercise

To review the structure and functions of major cellular components and to observe examples of human cells. To measure and compare the average cell's metabolic rate in individuals of different sizes (weight).

Learning Outcomes

After completing this exercise, you should be able to

1. Name and locate the components of a cell.
2. Differentiate the functions of cellular components.
3. Prepare a wet mount of cells lining the inside of the cheek; stain the cells; and identify the plasma (cell) membrane, nucleus, and cytoplasm.
4. Examine cells on prepared slides of human tissues and identify their major components.
5. Explain the relationship of oxygen consumption with an animal's total weight.
6. Explain the relationship of oxygen consumption per gram of weight.
7. Explain the relationship of oxygen consumption and metabolic rate.
8. Calculate surface area–to–volume ratio and explain its significance to temperature regulation.
9. Integrate the concepts of weight, oxygen consumption, and metabolism and explain their application to the human body.

Materials Needed

Animal cell model
Clean microscope slides
Coverslips
Flat toothpicks
Medicine dropper
Methylene blue (dilute) or iodine-potassium-iodide stain
Prepared microscope slides of human tissues
Compound light microscope

For Learning Extension Activities:
Single-edged razor blade
Plant materials such as leaves, soft stems, fruits, onion peel, and vegetables
Cultures of *Amoeba* and *Paramecium*

Pre-Lab

Carefully read the introductory material and examine the entire lab. Be familiar with the basic structures and functions of a cell from lecture or the textbook. Answer the pre-lab questions.

Pre-Lab Questions: Select the correct answer for each of the following questions:

1. Which of the following cellular structures is *not* easily visible with the compound light microscope?
   - **a.** nucleus
   - **b.** DNA
   - **c.** cytoplasm
   - **d.** plasma membrane
2. Which of the following cellular structures is located in the nucleus?
   - **a.** nucleolus
   - **b.** ribosomes
   - **c.** mitochondria
   - **d.** endoplasmic reticulum
3. The outer boundary of a cell is the
   - **a.** mitochondrial membrane.
   - **b.** nuclear envelope.
   - **c.** Golgi apparatus.
   - **d.** plasma membrane.

Safety

▶ Review all the safety guidelines inside the front cover.
▶ Clean laboratory surfaces before and after laboratory procedures.
▶ Wear disposable gloves for the wet-mount procedures of the cells lining the inside of the cheek.
▶ Work only with your own materials when preparing the slide of the cheek cells. Observe the same precautions as with all body fluids.
▶ Dispose of laboratory gloves, slides, coverslips, and toothpicks as instructed.
▶ Precautions should be taken to prevent cellular stains from contacting your clothes and skin.
▶ Wash your hands before leaving the laboratory.

4. Microtubules, intermediate filaments, and microfilaments are components of
  a. vesicles.　　　**b.** the Golgi apparatus.
  c. the cytoskeleton.　**d.** ribosomes.

5. Easily attainable living cells observed in the lab are from
  a. inside the cheek.　**b.** blood.
  c. hair.　　　　　　**d.** finger surface.

6. Cellular energy is called
  a. ER.　　　　　**b.** ATP.
  c. DNA.　　　　**d.** RNA.

7. The smooth ER possesses ribosomes.
  True _____　　　False _____

8. The nuclear envelope contains nuclear pores.
  True _____　　　False _____

Cells are the "building blocks" from which all parts of the human body are formed. Their arrangement and interactions result in the shape, organization, and construction of the body and are responsible for carrying on its life processes. Using a compound light microscope and the proper stain, one can easily see the **plasma (cell) membrane,** the **cytoplasm,** and the **nucleus.** The cytoplasm is composed of a clear fluid, the *cytosol,* and numerous *cytoplasmic organelles* that are suspended in the cytosol.

The plasma membrane, composed of phospholipids, glycolipids, and glycoproteins, represents the cell boundary and functions in various methods of membrane transport. The nucleus is surrounded by a nuclear envelope, and many of the cytoplasmic organelles have membrane boundaries similar to the plasma membrane. Movements of substances across these membranes can be by passive processes, involving kinetic energy or hydrostatic pressure, or active processes, using the cellular energy of ATP (adenosine triphosphate).

FIGURE 2.1　The structures of a composite cell. The structures are not drawn to scale.

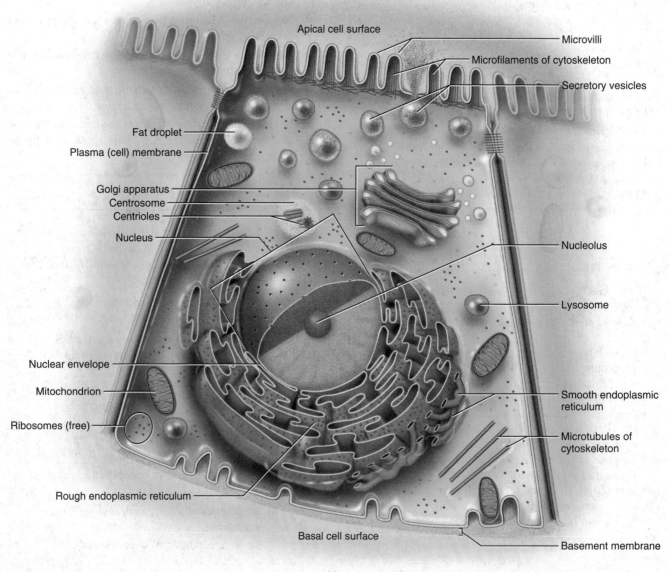

FIGURE 2.2 Structure of the plasma (cell) membrane.

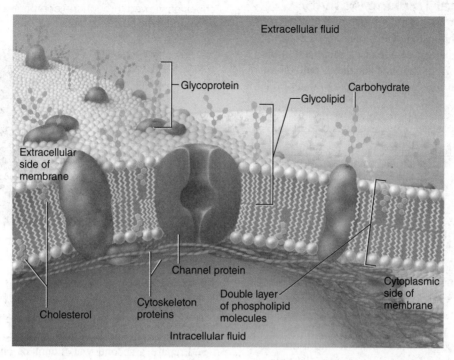

Metabolism refers to all the chemical reactions within cells that use or release energy. Is the metabolic rate of cells the same in a man who is six feet tall as in one who is five foot nine? Surprisingly, an understanding of temperature regulation is required to answer this question. *Temperature regulation* is accomplished through a balance of heat production and heat loss. Heat is produced by cells through metabolism (second law of thermodynamics: For every chemical reaction, some energy is always lost as heat). The greater the number of cells or the more metabolically active the cells, the more heat that is produced. Heat is lost through the surface of the skin. The larger the surface area of the skin, the more that is lost. For a given volume of cells, the metabolic rate of cells is established and maintained to offset the heat that is lost by the surface of the skin (within limits). The more heat that is lost at the skin, the more metabolically active the cells must be to produce the heat required to keep the body warm.

Procedure A—Cell Structure and Function

The boundary of the nucleus is a double-layered *nuclear envelope,* which has nuclear pores allowing the passage of genetic information. The nucleus contains fine strands of DNA (deoxyribonucleic acid) and structural and regulatory proteins called *chromatin,* which condenses to form chromosomes during cell divisions. The *nucleolus* is a nonmembranous structure composed of RNA (ribonucleic acid) and protein and is a site of ribosomal formation.

Cytoplasmic organelles provide for specialized metabolic functions. The *endoplasmic reticulum (ER)* has numerous canals that serve in transporting molecules as proteins throughout the cytoplasm. *Ribosomes* synthesize proteins and are located free in the cytoplasm or on the surface of the endoplasmic reticulum (rough ER). If the ER lacks the ribosomes on the surface, it is called smooth ER and does not serve as a region of protein synthesis. The *Golgi apparatus (complex),* composed of flattened membranous sacs, is the site for packaging glycoproteins for transport and secretion. *Mitochondria* possess a double membrane and provide the main location for the cellular energy production of ATP. Mitochondria are often referred to as the "powerhouse" of a cell because of the energy production. *Lysosomes* are membranous sacs that contain intracellular digestive enzymes for destroying debris and worn-out organelles. *Vesicles* are membranous sacs produced by the cell that contain substances for storage or transport, or form from pinching off pieces of the plasma membrane. A *cytoskeleton* contains microtubules, intermediate filaments, and microfilaments that support cellular structures within the cytoplasm and are involved in cellular movements.

1. Study figures 2.1 and 2.2.
2. Observe the animal cell model and identify its major structures.
3. Complete Part A of Laboratory Assessment 2.
4. Prepare a wet mount of cells lining the inside of the cheek. To do this, follow these steps:
   a. Gently scrape (force is not necessary and should be avoided) the inner lining of your cheek with the broad end of a flat toothpick.
   b. Stir the toothpick in a drop of water on a clean microscope slide and dispose of the toothpick as directed by your instructor.

Critical Thinking Activity

The cells lining the inside of the cheek are frequently removed for making observations of basic cell structure. The cells are from stratified squamous epithelium. Explain why these cells are used instead of outer body surface tissue.

_____

_____

_____

Learning Extension Activity

Investigate the microscopic structure of various plant materials. To do this, prepare tiny, thin slices of plant specimens, using a single-edged razor blade. _(Take care not to injure yourself with the blade.)_ Keep the slices in a container of water until you are ready to observe them. To observe a specimen, place it into a drop of water on a clean microscope slide and cover it with a coverslip. Use the microscope and view the specimen using low- and high-power magnifications. Observe near the edges where your section of tissue is most likely to be one cell thick. Add a drop of dilute methylene blue or iodine-potassium-iodide stain, and note if any additional structures become visible. How are the microscopic structures of the plant specimens similar to the human tissues you observed? _____

_____

How are they different? _____

_____

Learning Extension Activity

Prepare a wet mount of the _Amoeba_ and _Paramecium_ by putting a drop of culture on a clean glass slide. Gently cover with a clean coverslip. Observe the movements of the _Amoeba_ with pseudopodia and the _Paramecium_ with cilia. Try to locate cellular components such as the plasma (cell) membrane, nuclear envelope, nucleus, mitochondria, and contractile vacuoles. Describe the movement of the _Amoeba_.

_____

_____

Describe the movement of the _Paramecium_.

_____

_____

FIGURE 2.3 Stained cell lining the inside of the cheek as viewed through the compound light microscope using the high-power objective (400×).

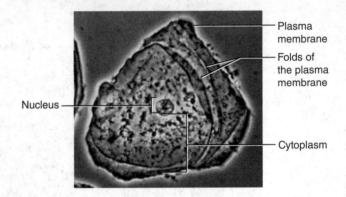

c. Cover the drop with a coverslip.
d. Observe the cheek cells by using the microscope. Compare your image with figure 2.3. To report what you observe, sketch a single cell in the space provided in Part B of the laboratory assessment.

5. Prepare a second wet mount of cheek cells, but this time, add a drop of dilute methylene blue or iodine-potassium-iodide stain to the cells. Cover the liquid with a coverslip and observe the cells with the microscope. Add to your sketch any additional structures you observe in the stained cells.

6. Answer the questions in Part B of the laboratory assessment.

7. Using the microscope, observe each of the prepared slides of human tissues. To report what you observe, sketch a single cell of each type in the space provided in Part C of the laboratory assessment.

8. Complete Parts C and D of the laboratory assessment.

NOTES

Name _____

Date _____

Section _____

The 🅐 corresponds to the indicated outcome(s) found at the beginning of the laboratory exercise.

Cell Structure and Function

Part A Assessments

1. Label the cellular structures in figure 2.4. 🅐

FIGURE 2.4   Label the indicated cellular structures of this composite cell.

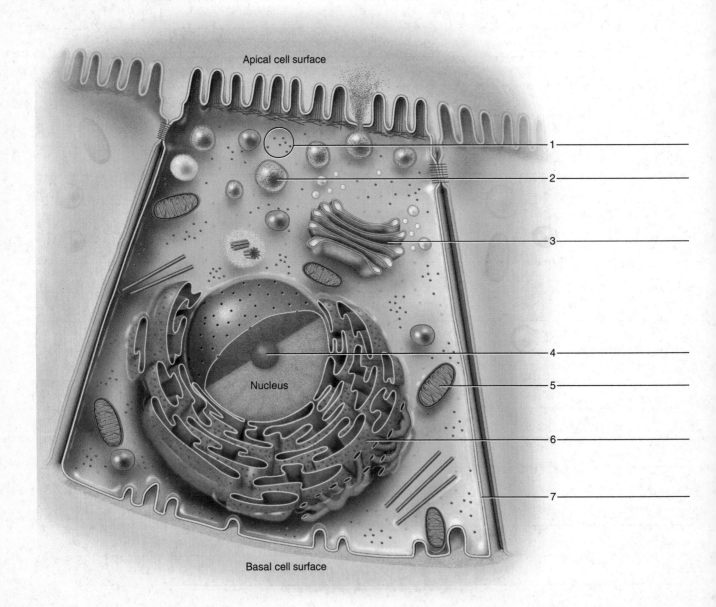

2. Match the cellular components in column A with the descriptions in column B. Place the letter of your choice in the space provided. 🔼2

Column A	Column B

Column A

a. Chromatin
b. Cytoplasm
c. Endoplasmic reticulum
d. Golgi apparatus (complex)
e. Lysosome
f. Microtubule
g. Mitochondrion
h. Nuclear envelope
i. Nucleolus
j. Nucleus
k. Ribosome
l. Vesicle

Column B

_____ 1. Loosely coiled fibers containing protein and DNA within nucleus

_____ 2. Location of ATP production for cellular energy

_____ 3. Small RNA-containing particles for the synthesis of proteins

_____ 4. Membranous sac formed by the pinching off of pieces of plasma membrane

_____ 5. Dense body of RNA and protein within the nucleus

_____ 6. Part of the cytoskeleton involved in cellular movement

_____ 7. Composed of membrane-bound canals for tubular transport throughout the cytoplasm

_____ 8. Occupies space between plasma membrane and nucleus

_____ 9. Flattened membranous sacs that package a secretion

_____ 10. Membranous sac that contains digestive enzymes

_____ 11. Separates nuclear contents from cytoplasm

_____ 12. Spherical organelle that contains chromatin and nucleolus

Part B   Assessments

Complete the following:

1. Sketch a single cheek cell. Label the cellular components you recognize. Add any additional structures observed to your sketch after staining was completed. (The circle represents the field of view through the microscope.) 🔼3

2. After comparing the wet mount and the stained cheek cells, describe the advantage gained by staining cells.

Magnification _____ ×

_____

_____

Part C Assessments

Complete the following:

1. Sketch a single cell of each type you observed in the prepared slides of human tissues. Name the tissue, indicate the magnification used, and label the cellular components you recognize.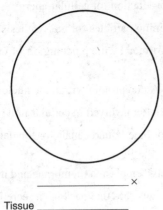

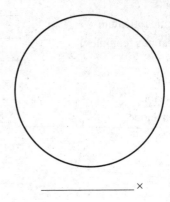

_____ ×

Tissue _____

_____ ×

Tissue _____

2. What do the various types of cells in these tissues have in common? _____

_____

3. What are the main differences you observed among these cells? _____

_____

_____

Part D Assessments

Electron micrographs represent extremely thin slices of cells. Each micrograph in figure 2.5 contains a section of a nucleus and some cytoplasm. Compare the organelles shown in these micrographs with organelles of the animal cell model and figure 2.1.

Identify the structures indicated by the arrows in figure 2.5. ⚠

1. _____

2. _____

3. _____

4. _____

5. _____

6. _____

7. _____

8. _____

9. _____

10. _____

FIGURE 2.5 Transmission electron micrographs of cellular components. The views are only portions of a cell. Magnifications: (a) 26,000×; (b) 10,000×. Identify the numbered cellular structures, using the terms provided.

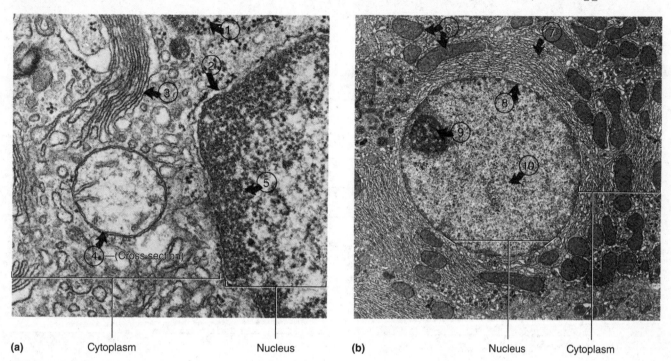

(a)   Cytoplasm   Nucleus   (b)   Nucleus   Cytoplasm

Terms:
Chromatin (use 2 times)
Endoplasmic reticulum
Golgi apparatus
Mitochondria
Mitochondrion (cross section)
Nuclear envelope (use 2 times)
Nucleolus
Ribosomes

Answer the following questions after observing the transmission electron micrographs in figure 2.5.

11. What cellular structures were visible in the transmission electron micrographs that were not apparent in the cells you observed using the microscope? _____

_____

_____

17

Movements Through Membranes

Purpose of the Exercise

To demonstrate some of the physical processes by which substances move through membranes.

⚠ Safety

- ▶ Clean laboratory surfaces before and after laboratory procedures.
- ▶ Wear disposable gloves when handling chemicals and animal blood.
- ▶ Wear safety glasses when using chemicals.
- ▶ Dispose of laboratory gloves and blood-contaminated items as instructed.
- ▶ Wash your hands before leaving the laboratory.

Materials Needed

For Procedure A— Diffusion:
Petri dish
White paper
Forceps

Potassium permanganate crystals
Millimeter ruler (thin and transparent)

For Procedure B— Osmosis:
Thistle tube
Molasses (or Karo dark corn syrup)
Ring stand and clamp
Beaker
Rubber band

Millimeter ruler
Selectively permeable (semipermeable) membrane (presoaked dialysis tubing of 1 5/16" or greater diameter)

For Procedure C— Hypertonic, Hypotonic, and Isotonic Solutions:
Test tubes
Marking pen
Test-tube rack
10 mL graduated cylinder
Medicine dropper

Uncoagulated animal blood
Distilled water
0.9% NaCl (aqueous solution)
3.0% NaCl (aqueous solution)
Clean microscope slides
Coverslips
Compound light microscope

For Procedure D— Filtration:
Glass funnel
Filter paper
Ring stand and ring
Beaker
Powdered charcoal or ground black pepper
1% glucose (aqueous solution)

1% starch (aqueous solution)
Test tubes
10 mL graduated cylinder
Water bath (boiling water)
Benedict's solution
Iodine-potassium-iodide solution
Medicine dropper

For Alternative Osmosis Activity:
Fresh chicken egg
Beaker

Laboratory balance
Spoon
Vinegar
Corn syrup (Karo)

Learning Outcomes

After completing this exercise, you should be able to

1. Show diffusion and identify examples of diffusion.
2. Interpret diffusion by preparing and explaining a graph.
3. Show osmosis and identify examples of osmosis.
4. Distinguish among hypertonic, hypotonic, and isotonic solutions and examine the effects of these solutions on animal cells.
5. Show filtration and identify examples of filtration.
6. Identify the percent transmittance of light through a blood sample placed in varying concentrations of NaCl.
7. Interpret the results to explain the relationship of percent transmittance and condition of the red blood cells (normal, shriveled or ruptured) in the blood sample.
8. Interpret the results to explain the condition of the red blood cells (normal, shriveled, or ruptured) to the relative concentration of NaCl solutions (isotonic, hypertonic, or hypotonic).
9. Relate plasma concentration and red blood cell integrity with events that occur in the human body.

Pre-Lab

Carefully read the introductory material and examine the entire lab. Be familiar with diffusion, osmosis, tonicity, and filtration from lecture or the textbook. Answer the pre-lab questions.

Pre-Lab Questions: Select the correct answer for each of the following questions:

1. Which of the following membrane movements is *not* by a passive process?
   - **a.** osmosis
   - **b.** diffusion
   - **c.** active transport
   - **d.** filtration

2. Osmosis is a membrane passage process pertaining to
   - **a.** glucose solutions.
   - **b.** hydrostatic pressure.
   - **c.** saline solutions.
   - **d.** water.

3. Which of the following represents an isotonic solution to human cells?
   - **a.** 0.9 % NaCl
   - **b.** 3 % NaCl
   - **c.** 3 % glucose
   - **d.** 0.9 % glucose

4. Filtration requires _____ for movements through membranes.
   - **a.** hydrostatic pressure
   - **b.** active transport
   - **c.** pinocytosis
   - **d.** phagocytosis

5. Which of the following does *not* influence the rate of diffusion?
   - **a.** concentration gradient
   - **b.** size of petri dish
   - **c.** molecular size
   - **d.** temperature

6. Dialysis tubing serves as a _____ for the osmosis experiment.
   - **a.** cell wall
   - **b.** permeable membrane
   - **c.** selectively permeable
   - **d.** barrier membrane

7. Passive membrane passage requires ATP.
   True _____ False _____

8. A cell will not swell or shrink when placed in an isotonic solution.
   True _____ False _____

A plasma membrane functions as a gateway through which chemical substances and small particles may enter or leave a cell. These substances move through the membrane by passive (physical) processes such as diffusion, osmosis, and filtration, or by active (physiological) processes such as active transport, phagocytosis, or pinocytosis. Passive processes do not require the cellular energy of ATP, but occur due to molecular motion or hydrostatic pressure; active processes utilize ATP.

This laboratory exercise contains examples of passive membrane transport through either living plasma membranes or artificial membranes. *Diffusion* is the random motion of molecules from an area of higher concentration toward an area of lower concentration. The rate of diffusion depends upon factors such as the concentration gradient, the molecular size, and the temperature. Facilitated diffusion is a carrier-mediated transport, which is not represented in this laboratory exercise.

A special case of passive membrane passage, called *osmosis,* occurs when water molecules diffuse through a selectively permeable membrane from an area of higher water concentration toward an area of lower water concentration. *Tonicity* refers to the osmotic pressure of a solution in relation to a cell. A solution is isotonic if the osmotic pressure is the same as the cell. A 5% glucose solution and a 0.9% normal saline solution serve as isotonic solutions to human cells. A hypertonic solution has a higher concentration of solutes than the cell; a hypotonic solution has a lower concentration of solutes than a cell. The osmosis of water will occur out of a cell when immersed in a hypertonic solution, and the cell will shrink (crenate). Conversely, the osmosis of water will occur into a cell when immersed in a hypotonic solution, and the cell will swell, and may burst (lyse).

Filtration is the movement of water and solutes through a selectively permeable membrane by hydrostatic pressure upon the membrane. The pressure gradient forces the water and solutes (filtrate) from the higher hydrostatic pressure area to a lower hydrostatic pressure area. For example, blood pressure provides the hydrostatic pressure for water and dissolved substances to move through capillary walls.

Procedure A—Diffusion

Simple diffusion can occur in living organisms as well as in nonliving systems involving gases, liquids, and solids. In this procedure, diffusion will be examined in water without cells.

1. To demonstrate diffusion, refer to figure 3.1 as you follow these steps:
   - **a.** Place a petri dish, half filled with water, on a piece of white paper that has a millimeter ruler positioned on the paper. Wait until the water surface is still. Allow approximately 3 minutes.
     Note: The petri dish should remain level. A second millimeter ruler may be needed under the petri dish as a shim to obtain a level amount of the water inside the petri dish.
   - **b.** Using forceps, place one crystal of potassium permanganate near the center of the petri dish and near the millimeter ruler (fig. 3.1).
   - **c.** Measure the radius of the purple circle at 1-minute intervals for 10 minutes and record the results in Part A of Laboratory Assessment 3.
2. Complete Part A of the assessment.

FIGURE 3.1 To demonstrate diffusion, place one crystal of potassium permanganate in the center of a petri dish containing water. Place the crystal near the millimeter ruler (positioned under the petri dish).

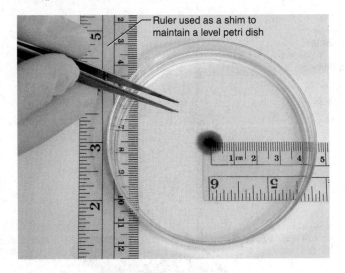

Ruler used as a shim to maintain a level petri dish

Learning Extension Activity

Repeat the demonstration of diffusion using a petri dish filled with ice-cold water and a second dish filled with very hot water. At the same moment, add a crystal of potassium permanganate to each dish and observe the circle as before. What difference do you note in the rate of diffusion in the two dishes? How do you explain this difference? _____

_____

_____

_____

_____

Procedure B—Osmosis

For the osmosis experiment, an artificial membrane will serve as a model for an actual plasma membrane. A layer of dialysis tubing that has been soaked for 30 minutes can easily be cut open because it becomes pliable. This membrane model works as a selectively (semipermeable; differentially) permeable membrane.

When water moves across a selectively permeable membrane down its concentration gradient, it is termed osmosis. Solute particles will also move down their concentration gradients across a membrane if the solute will pass through the membrane. However, if all the solute particles are nonpenetrating because they are too large to pass across a selectively permeable membrane, only water will move across the selectively permeable membrane. This will result in changes in the volume of water on a certain side of the membrane. This will create a volume change in relation to a cell or models used as simulated cells.

1. To demonstrate osmosis, refer to figure 3.2 as you follow these steps:
    a. One person plugs the tube end of a thistle tube with a finger.
    b. Another person then fills the bulb with molasses until it is about to overflow at the top of the bulb. Air remains trapped in the stem.
    c. Cover the bulb opening with a single-thickness piece of moist selectively permeable membrane.
    d. Tightly secure the membrane in place with several wrappings of a rubber band.
    e. Immerse the bulb end of the tube in a beaker of water. If leaks are noted, repeat the procedures.
    f. Support the upright portion of the tube with a clamp on a ring stand. Folded paper under the clamp will protect the thistle tube stem from breakage.
    g. Mark the meniscus level of the molasses in the tube. *Note:* The best results will occur if the mark of the molasses is a short distance up the stem of the thistle tube when the experiment starts.
    h. Measure the level changes after 10 minutes and 30 minutes and record the results in Part B of the laboratory assessment.

2. Complete Part B of the laboratory assessment.

Alternative Activity

Eggshell membranes possess selectively permeable properties. To demonstrate osmosis using a natural membrane, soak a fresh chicken egg in vinegar for about 24 hours to remove the shell. Use a spoon to carefully handle the delicate egg. Place the egg in a hypertonic solution (corn syrup) for about 24 hours. Remove the egg, rinse it, and using a laboratory balance, weigh the egg to establish a baseline weight. Place the egg in a hypotonic solution (distilled water). Remove the egg and weigh it every 15 minutes for an elapsed time of 75 minutes. Explain any weight changes that were noted during this experiment.

_____

_____

_____

_____

_____

FIGURE 3.2 (a) Fill the bulb of the thistle tube with molasses; (b) tightly secure a piece of selectively permeable (semipermeable) membrane over the bulb opening; and (c) immerse the bulb in a beaker of water. Note: These procedures require the participation of two people.

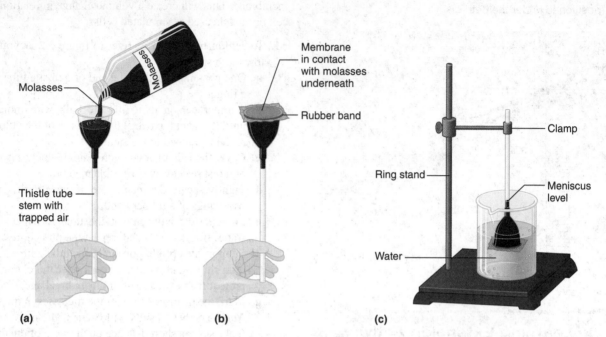

(a)   (b)   (c)

Procedure C—Hypertonic, Hypotonic, and Isotonic Solutions

Applications of volume changes to cells from osmosis are addressed in this procedure. If a cell is placed in an isotonic solution with the same water and solute concentration, there will be no net osmosis and the cell will retain its normal size. If a cell is placed in a hypertonic solution, the cell will lose water and shrink; if a cell is in a hypotonic solution it will swell and burst.

1. To demonstrate the effect of hypertonic, hypotonic, and isotonic solutions on animal cells, follow these steps:
   a. Place three test tubes in a rack and mark them *tube 1, tube 2,* and *tube 3. (Note:* One set of tubes can be used to supply test samples for the entire class.)
   b. Using 10 mL graduated cylinders, add 3 mL of distilled water to tube 1; add 3 mL of 0.9% NaCl to tube 2; and add 3 mL of 3.0% NaCl to tube 3.
   c. Place three drops of fresh, uncoagulated animal blood into each of the tubes, and gently mix the blood with the solutions. Wait 5 minutes.
   d. Using three separate medicine droppers, remove a drop from each tube and place the drops on three separate microscope slides marked *1, 2,* and *3.*
   e. Cover the drops with coverslips and observe the blood cells, using the high power of the microscope.
2. Complete Part C of the laboratory assessment.

 Alternative Activity

Various substitutes for blood can be used for Procedure C. Onion, cucumber, or cells lining the inside of the cheek represent three possible options.

Procedure D—Filtration

Filtration is another example of passive membrane passage. Unlike the previous experiments, hydrostatic pressure provides the mechanism for movements across a membrane instead of molecular motion. Water and solutes that pass through the membrane move simultaneously in the same direction. Solutes that are too large to cross the membrane remain on their original side of the membrane. Filtration occurs faster when the hydrostatic pressures increase.

1. To demonstrate filtration, follow these steps:
   a. Place a glass funnel in the ring of a ring stand over an empty beaker. Fold a piece of filter paper in half and then in half again. Open one thickness of the filter paper to form a cone. Wet the cone, and place it in the funnel. The filter paper is used to demonstrate how movement across membranes is limited by the size of the molecules, but it does not represent a working model of biological membranes.
   b. Prepare a mixture of 5 cc (approximately 1 teaspoon) powdered charcoal (or ground black pepper) and equal amounts of 1% glucose solution and 1% starch solution in a beaker. Pour some of the mixture into the funnel until it nearly reaches the top of the filter-paper

FIGURE 3.3 Apparatus used to illustrate filtration.

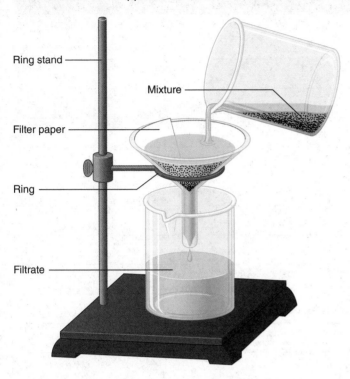

Ring stand

Mixture

Filter paper

Ring

Filtrate

cone. Care should be taken to prevent the mixture from spilling over the top of the filter paper. Collect the filtrate in the beaker below the funnel (fig. 3.3).

c. Test some of the filtrate in the beaker for the presence of glucose. To do this, place 1 mL of filtrate in a clean test tube and add 1 mL of Benedict's solution. Place the test tube in a water bath of boiling water for 2 minutes and then allow the liquid to cool slowly. If the color of the solution changes to green, yellow, or red, glucose is present (fig. 3.4).

d. Test some of the filtrate in the beaker for the presence of starch. To do this, place a few drops of filtrate in a test tube and add one drop of iodine-potassium-iodide solution. If the color of the solution changes to blue-black, starch is present.

e. Observe any charcoal in the filtrate.

2. Complete Part D of the laboratory assessment.

FIGURE 3.4 Heat the filtrate and Benedict's solution in a boiling water bath for 2 minutes.

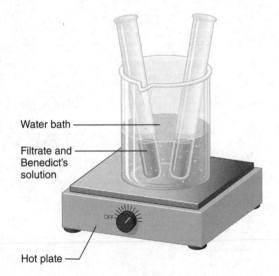

Water bath

Filtrate and Benedict's solution

Hot plate

Name _____

Date _____

Section _____

The 🄰 corresponds to the indicated outcome(s) found at the beginning of the laboratory exercise.

Movements Through Membranes

Part A Assessments

Complete the following:

1. Enter data for changes in the movement of the potassium permanganate. 🄰

Elapsed Time	Radius of Purple Circle in Millimeters
Initial	_____
1 minute	_____
2 minutes	_____
3 minutes	_____
4 minutes	_____
5 minutes	_____
6 minutes	_____
7 minutes	_____
8 minutes	_____
9 minutes	_____
10 minutes	_____

2. Prepare a graph that illustrates the diffusion distance of potassium permanganate in 10 minutes. 🄰

3. Explain your graph. 🄰 _____

_____

_____

_____

_____

_____

_____

_____

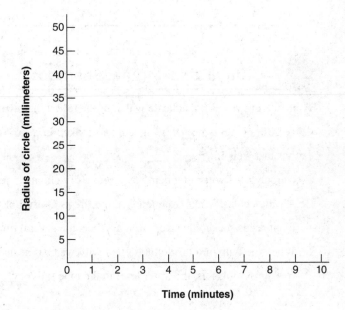

25

4. Define *diffusion.* _____

_____

_____

_____

Critical Thinking Assessment

By answering yes or no, indicate which of the following provides an example of diffusion. ⚠️

1. A perfume bottle is opened, and soon the odor can be sensed in all parts of the room. _____

2. A sugar cube is dropped into a cup of hot water, and, without being stirred, all of the liquid becomes sweet tasting. _____

3. Water molecules move from a faucet through a garden hose when the faucet is turned on. _____

4. A person blows air molecules into a balloon by forcefully exhaling. _____

5. A crystal of blue copper sulfate is placed in a test tube of water. The next day, the solid is gone, but the water is evenly colored. _____

Part B   Assessments

Complete the following:

1. What was the change in the level of molasses in 10 minutes?_____

2. What was the change in the level of molasses in 30 minutes?_____

3. How do you explain this change? ⚠️ _____

_____

4. Define *osmosis.* _____

_____

Critical Thinking Assessment

By answering yes or no, indicate which of the following involves osmosis. ⚠️

1. A fresh potato is peeled, weighed, and soaked in a strong salt solution. The next day, it is discovered that the potato has lost weight. _____

2. Garden grass wilts after being exposed to dry chemical fertilizer. _____

3. Air molecules escape from a punctured tire as a result of high pressure inside. _____

4. Plant seeds soaked in water swell and become several times as large as before soaking. _____

5. When the bulb of a thistle tube filled with water is sealed by a selectively permeable membrane and submerged in a beaker of molasses, the water level in the tube falls. _____

Part C Assessments

Complete the following:

1. In the spaces, sketch a few blood cells from each of the test tubes and indicate the magnification.

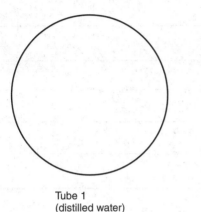

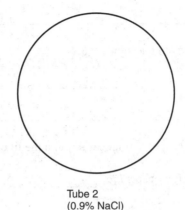

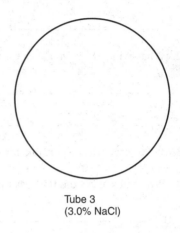

Tube 1
(distilled water)

Tube 2
(0.9% NaCl)

Tube 3
(3.0% NaCl)

_____ ×

_____ ×

_____ ×

2. Based on your results, which tube contained a solution hypertonic to the blood cells? A _____

   Give the reason for your answer. A _____

   _____

3. Which tube contained a solution hypotonic to the blood cells? A _____

   Give the reason for your answer. A _____

   _____

4. Which tube contained a solution isotonic to the blood cells? A _____

   Give the reason for your answer. A _____

   _____

5. Observe the RBCs shown in figure 3.5. Select the solutions in which the cells were placed to illustrate the effects of tonicity on cells.

FIGURE 3.5 Three blood cells placed in three different solutions: distilled water, 0.9% NaCl, and 3% NaCl. Select the solutions in which the cells were placed, using the terms provided. A

Terms:
Hypertonic
Hypotonic
Isotonic

(a) _____

(b) _____

(c) _____

27

Part D   Assessments

Complete the following:

1. Which of the substances in the mixture you prepared passed through the filter paper into the filtrate? ⑤ _____

   _____

2. What evidence do you have for your answer to question 1? ⑤ _____

   _____

   _____

3. What force was responsible for the movement of substances through the filter paper? ⑤ _____

4. What substances did not pass through the filter paper? ⑤ _____

   _____

5. What factor prevented these substances from passing through? ⑤ _____

   _____

6. Define *filtration.* _____

   _____

Critical Thinking Assessment

By answering yes or no, indicate which of the following involves filtration. ⑤

1. Oxygen molecules move into a cell and carbon dioxide molecules leave a cell because of differences in the concentrations of these substances on either side of the plasma membrane._____

2. Blood pressure forces water molecules from the blood outward through the thin wall of a blood capillary. _____

3. Urine is forced from the urinary bladder through the tubular urethra by muscular contractions._____

4. Air molecules enter the lungs through the airways when air pressure is greater outside these organs than inside._____

5. Coffee is made using a coffeemaker (not instant)._____

Critical Thinking Assessment

A patient is given an IV (intravenous) of deionized water (water with no solutes). If a large enough amount of fluid is administered, predict the effects. Choose one of the words in parentheses for each blank to complete the sentence. Plasma becomes

_____ (hypotonic/hypertonic) to the red blood cells.

_____ Water moves from the (plasma or red blood cells)

_____ to the (plasma or red blood cells)

_____ causing the red blood cells to (shrivel or swell).

A person drinks ocean salt water (3.5% salt). If the person drinks a large amount of the salt water (the salt water is absorbed into the blood), the plasma becomes

_____ (hypotonic/hypertonic) to the red blood cells.

_____ Water moves from the (plasma or red blood cells)

_____ to the (plasma or red blood cells)

_____ causing the red blood cells to (shrivel or swell).

What would you predict if a patient was administered an IV containing a physiological saline solution (0.9% sodium chloride)? _____

_____

29

NOTES

Muscle and Nervous Tissues

Purpose of the Exercise

To review the characteristics of muscle and nervous tissues and to observe examples of these tissues.

Materials Needed

Compound light microscope
Prepared slides of the following:
    Skeletal muscle tissue
    Smooth muscle tissue
    Cardiac muscle tissue
    Nervous tissue (spinal cord smear and/or
      cerebellum)
For Learning Extension Activity:
Colored pencils

Learning Outcomes

After completing this exercise, you should be able to

1. Sketch and label the characteristics of muscle tissues and nervous tissues that you were able to observe.

2. Differentiate the special characteristics of each type of muscle tissue and nervous tissue.

3. Indicate a location and function of each type of muscle tissue and nervous tissue.

Pre-Lab

Carefully read the introductory material and examine the entire lab. Be familiar with muscle tissues and nervous tissue from lecture or the textbook. Answer the pre-lab questions.

Pre-Lab Questions: Select the correct answer for each of the following questions:

1. Which muscle tissue is under conscious control?
     a. skeletal   **b.** smooth
     c. cardiac   **d.** stomach

2. Which of the following organs lacks smooth muscle?
     a. blood vessels   **b.** stomach
     c. iris   **d.** heart

3. Which muscle tissue lacks striations?
     a. cardiac   **b.** skeletal
     c. smooth   **d.** voluntary

4. Which of the following conduct action potentials?
     a. smooth muscle cells   **b.** neurons
     c. neuroglia   **d.** skeletal muscle cells

5. Muscles of facial expression are
     a. smooth muscles.   **b.** skeletal muscles.
     c. involuntary muscles.   **d.** cardiac muscles.

6. Which muscle tissue is multinucleated?
     a. skeletal   **b.** smooth
     c. stomach   **d.** cardiac

7. Intercalated discs represent the junction where heart muscle cells fit together.
   True _____ False _____

8. Both cardiac and skeletal muscle cells are voluntary.
   True _____ False _____

Muscle tissues are characterized by the presence of elongated cells, often called muscle fibers, that can contract to create movements. Many of our muscles are attached to the skeleton, but muscles are also components of many of our internal organs. During muscle contractions, considerable body heat is generated to help maintain our body temperature. Because more heat is generated than is needed to maintain our body temperature, much of the heat is dissipated from our body through the skin.

The three types of muscle tissues are *skeletal, smooth,* and *cardiac.* Skeletal muscles are under our conscious control and are considered voluntary. Although most skeletal muscles are attached to bones via tendons, the tongue, facial muscles, and voluntary sphincters do not attach directly to the bone. Functions include body movements, maintaining posture, breathing, speaking, controlling waste eliminations, and protection. Skeletal muscles also aid the movement of lymph and venous blood during muscle contractions. Smooth muscle is considered involuntary and is located in many visceral organs, the iris, blood vessels, respiratory tubes, and attached to hair follicles. Functions include the motions of visceral organs (peristalsis), controlling pupil size, blood flow, and airflow, and creating "goose bumps" if we are too cold or frightened. Cardiac muscle is located only in the heart wall. It is considered involuntary and functions to pump blood.

Nervous tissues occur in the brain, spinal cord, and peripheral nerves. The tissue consists of two cell types: *neurons* and *neuroglia.* Neurons, also called nerve cells, contain a cell body with the nucleus and most of the cytoplasm, and cellular processes that extend from the cell body. Cellular processes include one to many dendrites and a single axon (nerve fiber). Neurons are considered excitable cells because they can generate signals called action potentials (nerve impulses) along the neuron to another neuron or a muscle or gland. Neuroglia (glial cells) of various types are more abundant than neurons; they cannot conduct nerve impulses, but they have important supportive and protective functions for neurons. Additional study of nervous tissue will be found in Laboratory Exercise 14.

Procedure—Muscle and Nervous Tissues

1. Using the microscope, observe each of the types of muscle tissues on the prepared slides. Look for the special features of each type. Compare your prepared slides of muscle tissues to the micrographs in figure 4.1

FIGURE 4.1 Micrographs of muscle and nervous tissues.

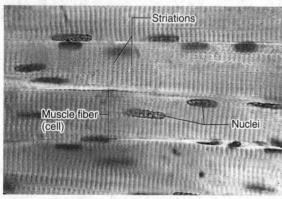

(a) Skeletal muscle (from leg) (400×)

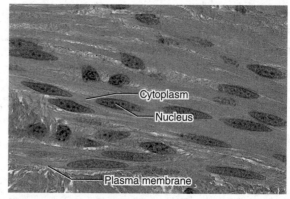

(b) Smooth muscle (from small intestine) (1,000×)

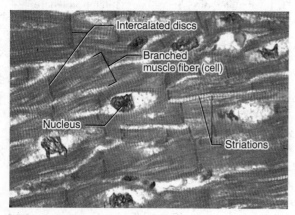

(c) Cardiac muscle (from heart) (400×)

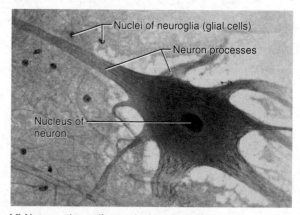

(d) Nervous tissue (from spinal cord) (400×)

and the characteristics of each specific tissue in tables 4.1 and 4.2.

2. As you observe each type of muscle tissue, prepare a labeled sketch of a representative portion of the tissue in Part A of Laboratory Assessment 4.
3. Observe the prepared slide of nervous tissue and identify neurons (nerve cells), neuron cellular processes, and neuroglia. Compare your prepared slide of nervous tissue to the micrograph in figure 4.1 and the characteristics in table 4.1.
4. Prepare a labeled sketch of nervous tissue in Part A of the laboratory assessment.
5. Test your ability to recognize each of these muscle and nervous tissues by having your laboratory partner select a slide, cover its label, and focus the microscope on this tissue. Then see if you correctly identify the tissue.
6. Complete Part B of the laboratory assessment.

TABLE 4.1 Muscle and Nervous Tissues, Descriptions, Functions, and Representative Locations

Tissue Type	Descriptions	Functions	Representative Locations
Skeletal muscle	Long, threadlike cells; striated; many nuclei near plasma membrane	Voluntary movements of skeletal parts; facial expressions	Muscles usually attached to bones
Smooth muscle	Shorter spindle-shaped cells; single central nucleus	Involuntary movements of internal organs	Walls of hollow internal organs
Cardiac muscle	Branched cells; striated; single nucleus (usually)	Heart contractions to pump blood; involuntary	Heart walls
Nervous tissue	Neurons with long cellular processes; neuroglia smaller and variable	Sensory reception and conduction of action potentials; neuroglia supportive	Brain, spinal cord, and peripheral nerves

TABLE 4.2 Muscle Tissue Characteristics

Characteristic	Skeletal Muscle	Smooth Muscle	Cardiac Muscle
Appearance of cells	Unbranched and relatively parallel	Spindle-shaped	Branched and connected in complex networks
Striations (pattern of alternating dark and light bands across cells)	Present and obvious	Absent	Present but faint
Nucleus	Multinucleated	Uninucleated	Uninucleated (usually)
Intercalated discs (junction where cells fit together)	Absent	Absent	Present
Control	Voluntary	Involuntary	Involuntary

Laboratory Assessment

4

Name _____

Date _____

Section _____

The Ⓐ corresponds to the indicated outcome(s) found at the beginning of the laboratory exercise.

Muscle and Nervous Tissues

Part A Assessments

In the space that follows, sketch a few cells or fibers of each of the three types of muscle tissues and of nervous tissue as they appear through the microscope. For each sketch, label the major structures of the cells or fibers, indicate the magnification used, write an example of a location in the body, and provide a function. Ⓐ1 Ⓐ2 Ⓐ3

Skeletal muscle tissue (_____×)

Location: _____

Function: _____

Smooth muscle tissue (_____×)

Location: _____

Function: _____

Cardiac muscle tissue (_____×)

Location: _____

Function: _____

Nervous tissue (_____×)

Location: _____

Function: _____

Learning Extension Activity

Use colored pencils to differentiate various cellular structures in Part A.

Part B Assessments

Match the tissues in column A with the characteristics in column B. Place the letter of your choice in the space provided. (Some answers may be used more than once.) 🔺 🔺

Column A	Column B
a. Cardiac muscle	_____ **1.** Coordinates, regulates, and integrates body functions
b. Nervous tissue	_____ **2.** Contains intercalated discs
c. Skeletal muscle	_____ **3.** Muscle that lacks striations
d. Smooth muscle	_____ **4.** Striated and involuntary
	_____ **5.** Striated and voluntary
	_____ **6.** Contains neurons and neuroglia
	_____ **7.** Muscle attached to bones
	_____ **8.** Muscle that composes heart
	_____ **9.** Moves food through the digestive tract
	_____ **10.** Transmits impulses along cellular processes
	_____ **11.** Muscle under conscious control
	_____ **12.** Muscle of blood vessels and iris

Integumentary System

Purpose of the Exercise

To observe the structures and tissues of the integumentary system and to review the functions of these parts.

Materials Needed

Skin model
Hand magnifier or dissecting microscope
Forceps
Microscope slide and coverslip
Compound light microscope
Prepared microscope slide of human scalp or axilla
Prepared slide of dark (heavily pigmented) human skin
Prepared slide of thick skin (plantar or palmar)

For Learning Extension Activity:
Tattoo slide
Stereomicroscope (dissecting microscope)

Learning Outcomes

After completing this exercise, you should be able to

1. Locate the structures of the integumentary system.
2. Describe the major functions of these structures.
3. Distinguish the locations and tissues among epidermis, dermis, and the hypodermis.
4. Sketch the layers of the skin and associated structures observed on the prepared slide.

Pre-Lab

Carefully read the introductory material and examine the entire lab. Be familiar with skin layers and accessory structures of the skin from lecture or the textbook. Answer the pre-lab questions.

Pre-Lab Questions: Select the correct answer for each of the following questions:

1. Which of the following is *not* a function of the integumentary system?
   a. protection
   b. excrete small amounts of waste
   c. movement
   d. aid in regulating body temperature

2. The two distinct skin layers are the
   a. epidermis and dermis.
   b. hypodermis and dermis.
   c. hypodermis and epidermis.
   d. dermis and hypodermis.

3. Apocrine sweat glands are located in _____ regions of the body.
   a. forehead   b. axillary and genital
   c. palmar   d. plantar

4. The hypodermis is composed of _____ tissues.
   a. adipose and stratified squamous epithelial
   b. areolar and dense irregular connective
   c. stratified squamous epithelial and adipose
   d. areolar and adipose connective

5. The _____ layer of the epidermis is only present in thick skin.
   a. stratum corneum   b. stratum lucidum
   c. stratum spinosum   d. stratum granulosum

6. Frequent cell division occurs in the _____ of the epidermis.
   a. stratum corneum   b. stratum spinosum
   c. stratum granulosum   d. stratum basale

7. The greatest concentration of melanin is in the dermis.
   True _____   False _____

8. Thick skin of the palms and soles contains five strata of the epidermis.
   True _____   False _____

The integumentary system includes the skin, hair, nails, sebaceous (oil) glands, and sweat (sudoriferous) glands. These structures provide a protective covering for deeper tissues, aid in regulating body temperature, retard water loss, house sensory receptors, synthesize various chemicals, and excrete small quantities of wastes.

The skin consists of two distinct layers. The outer layer, the *epidermis,* consists of stratified squamous epithelium. The inner layer, the *dermis,* consists of a superficial papillary region of areolar connective tissue and a thicker and deeper reticular region of dense irregular connective tissue. Beneath the dermis is the hypodermis (subcutaneous layer; superficial fascia) composed of adipose and areolar connective tissues. The hypodermis is not considered a true layer of the skin.

Accessory structures of the skin include nails, hair follicles, and skin glands. The hair, which grows through a depression from the epidermis, possesses a hair papilla at the base of the hair which contains a network of capillaries that supply the nutrients for cell divisions for hair growth within the hair bulb. As the cells of the hair are forced toward the surface of the body, they become keratinized and pigmented and die. Attached to the follicle is the arrector pili muscle that can pull the hair to a more upright position, causing goose bumps when experiencing cold temperatures or fear. A sebaceous gland secretes an oily sebum into the hair follicles, which keeps the hair and epidermal surface pliable and somewhat waterproof.

Sweat glands (sudoriferous glands) are distributed over most regions of the body and consist of two types of glands. The widespread eccrine sweat glands are most numerous on the palms, soles of the feet, and the forehead. Their ducts open to the surface at a sweat pore. Their secretions increase during hot days, physical exercise, and stress; they serve an excretory function and can help prevent our body temperature from overheating. The apocrine sweat glands are most abundant in the axillary and genital regions. Apocrine sweat ducts open into the hair follicles and become active at puberty. Their secretions increase during stress and pain and have little influence on thermoregulation.

Procedure—Integumentary System

In this procedure you will use a skin model and make comparisons to the figures in the lab manual to locate the layers and accessory structures of the skin. Hair structures will be observed from your own body with different magnifications and then compared to additional detail using prepared slides. Several micrographs of different magnifications are provided of vertical sections of the skin layers and accessory structures. Use a combination of all of the micrographs as you observe the skin slides available in your laboratory.

1. Use figures 5.1 and 5.2 and locate as many of these structures as possible on a skin model.
2. Use figure 5.3 as a guide to locate the specific epidermal layers (strata) on a skin model. Note the locations and descriptions from table 5.1.

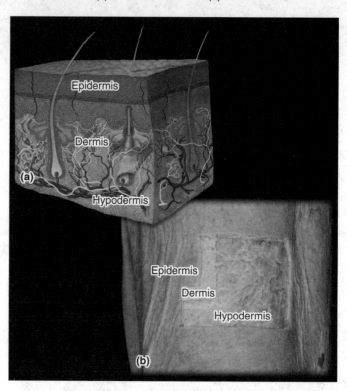

FIGURE 5.1 Layers of skin and deeper hypodermis indicated on (a) an illustration and (b) cadaver skin.

3. Use the hand magnifier or dissecting microscope and proceed as follows:
   a. Observe the skin, hair, and nails on your hand.
   b. Compare the type and distribution of hairs on the front and back of your forearm.
4. Use low-power magnification of the compound light microscope and proceed as follows:
   a. Pull out a single hair with forceps and mount it on a microscope slide under a coverslip.
   b. Observe the root and shaft of the hair and note the scalelike parts that make up the shaft.
5. Complete Parts A and B of Laboratory Assessment 5.
6. As vertical sections of human skin are observed, remember that the lenses of the microscope invert and reverse images. It is important to orient the position of the epidermis, dermis, and hypodermis layers using scan magnification before continuing with additional observations. Compare all of your skin observations to the various micrographs in figure 5.4. Use low-power magnification of the compound light microscope and proceed as follows:
   a. Observe the prepared slide of human scalp or axilla.
   b. Locate the epidermis, dermis, and hypodermis; a hair follicle; an arrector pili muscle; a sebaceous gland; and a sweat gland.
   c. Focus on the epidermis with high power and locate the stratum corneum, stratum granulosum, stratum spinosum, and stratum basale. Note how the shapes of the cells in these layers differ as described in table 5.1.

FIGURE 5.2 Vertical section of the skin and hypodermis (subcutaneous layer).

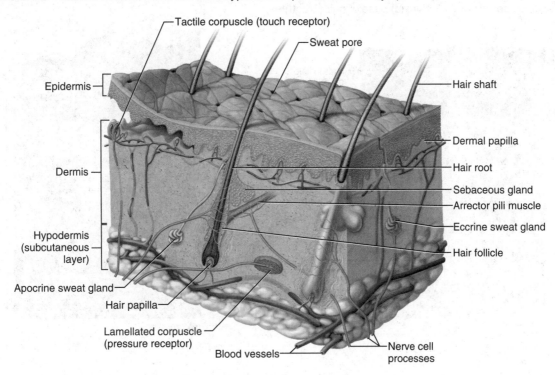

FIGURE 5.3 Epidermal layers in this section of thick skin from the fingertip (400×).

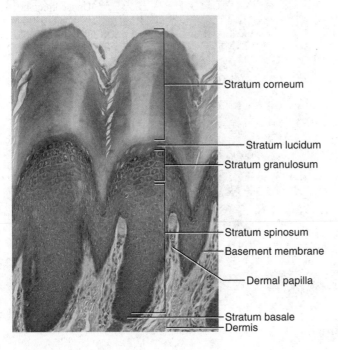

TABLE 5.1 Layers of the Epidermis

Layer	Location	Descriptions
Stratum corneum	Most superficial layer	Many layers of keratinized, dead epithelial cells; appear scaly and flattened; resists water loss, absorption, and abrasion
Stratum lucidum	Between stratum corneum and stratum granulosum on soles and palms of thick skin	Cells appear clear; nuclei, organelles, and plasma membranes no longer visible
Stratum granulosum	Beneath the stratum corneum (or stratum lucidum of thick skin)	Three to five layers of flattened granular cells; contain shrunken fibers of keratin and shriveled nuclei
Stratum spinosum	Beneath the stratum granulosum	Many layers of cells with centrally located, large, oval nuclei; develop fibers of keratin; cells becoming flattened in superficial portion
Stratum basale	Deepest layer	A single row of cuboidal or columnar cells; layer also includes melanocytes; frequent cell division; some cells become parts of more superficial layers

FIGURE 5.4 Features of human skin are indicated in these micrographs: (a) and (b) various structures of epidermis and dermis; (c) epidermis of dark skin; (d) base of hair structures.

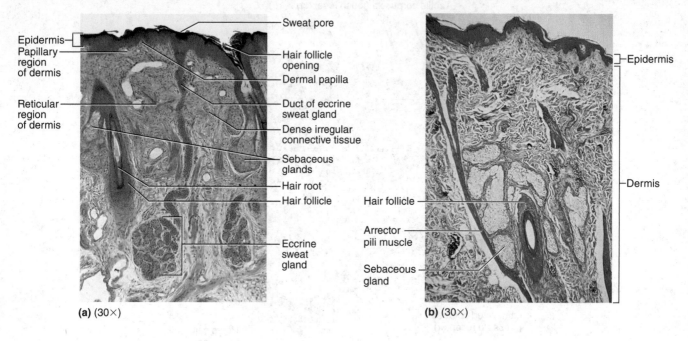

(a) (30×)

- Epidermis
- Papillary region of dermis
- Reticular region of dermis
- Sweat pore
- Hair follicle opening
- Dermal papilla
- Duct of eccrine sweat gland
- Dense irregular connective tissue
- Sebaceous glands
- Hair root
- Hair follicle
- Eccrine sweat gland

(b) (30×)

- Epidermis
- Dermis
- Hair follicle
- Arrector pili muscle
- Sebaceous gland

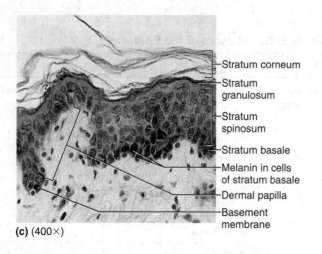

(c) (400×)

- Stratum corneum
- Stratum granulosum
- Stratum spinosum
- Stratum basale
- Melanin in cells of stratum basale
- Dermal papilla
- Basement membrane

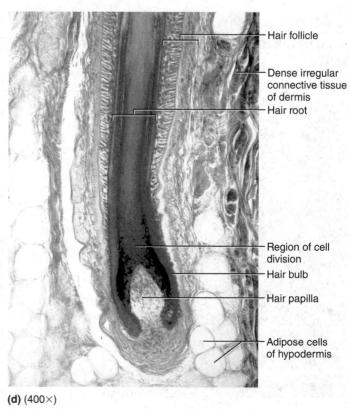

(d) (400×)

- Hair follicle
- Dense irregular connective tissue of dermis
- Hair root
- Region of cell division
- Hair bulb
- Hair papilla
- Adipose cells of hypodermis

9. Complete Part C of the laboratory assessment.
10. Using low-power magnification, locate a hair follicle sectioned longitudinally through its bulblike base. Also locate a sebaceous gland close to the follicle and find a sweat gland (fig. 5.4). Observe the detailed structure of these parts with high-power magnification.
11. Complete Parts D and E of the laboratory assessment.

Critical Thinking Activity

Explain the advantage for melanin granules being located in the deep layers of the epidermis, and not the dermis or deeper hypodermis.

_____

_____

_____

_____

_____

_____

d. Observe the dense irregular connective tissue that makes up the bulk of the dermis (reticular region).

e. Observe the adipose tissue that composes most of the hypodermis (subcutaneous layer).

7. Observe the prepared slide of dark (heavily pigmented) human skin with low-power magnification. Note that the pigment is most abundant in the deepest layers of the epidermis. Focus on this region with the high-power objective. The pigment-producing cells, or melanocytes, are located among the stratum basale cells. Some melanin is retained within some cells of the stratum spinosum as cells are forced closer to the surface of the skin. Differences in skin color are primarily due to the quantity of the pigment *melanin* produced by these cells. Exposure to ultraviolet (UV) rays of sunlight can increase the amount of melanin produced, causing a suntan. Melanin absorbs the UV radiation, which helps to protect the nuclei of cells.

Learning Extension Activity

Observe a vertical section of human skin through a tattoo, using low-power magnification. Note the location of the dispersed ink granules within the upper portion of the dermis. From a thin vertical section of a tattoo, it is not possible to determine the figure or word of the entire tattoo as seen on the surface of the skin. Compare this to the location of melanin granules found in dark (heavily pigmented) skin. Suggest reasons why a tattoo is permanent and a suntan is not.

_____

_____

_____

_____

_____

_____

Laboratory Assessment

5

Name _____

Date _____

Section _____

The Ⓐ corresponds to the indicated outcome(s) found at the beginning of the laboratory exercise.

Integumentary System

Part A Assessments

1. Label the structures indicated in figure 5.5. Ⓐ1

FIGURE 5.5   Label the features of the skin.

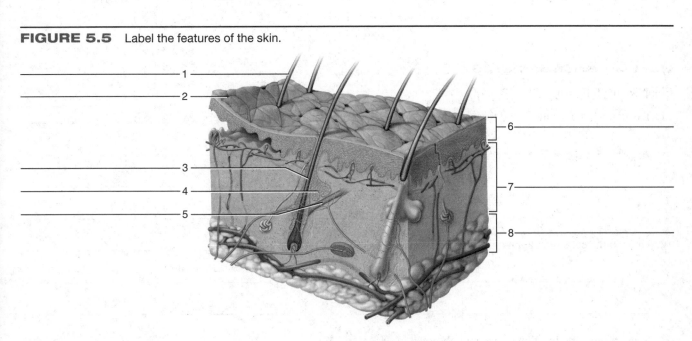

2. Match the structures in column A with the description and functions in column B. Place the letter of your choice in the space provided. Ⓐ1 Ⓐ2

Column A	Column B
a. Apocrine sweat gland	_____ **1.** An oily secretion that helps to waterproof body surface
b. Arrector pili muscle	_____ **2.** Outermost layer of epidermis
c. Dermis	_____ **3.** Become active at puberty
d. Eccrine (merocrine) sweat gland	_____ **4.** Epidermal pigment
e. Epidermis	_____ **5.** Inner layer of skin
f. Hair follicle	_____ **6.** Responds to elevated body temperature
g. Keratin	_____ **7.** General name of entire superficial layer of the skin
h. Melanin	_____ **8.** Gland that secretes an oily substance
i. Sebaceous gland	_____ **9.** Hard protein of nails and hair
j. Sebum	_____ **10.** Cell division and deepest layer of epidermis
k. Stratum basale	_____ **11.** Tubelike part that contains the root of the hair
l. Stratum corneum	_____ **12.** Causes hair to stand on end and goose bumps to appear

Part B Assessments

Complete the following:

1. How does the skin of your palm differ from that on the back (posterior) of your hand? _____

   _____

   _____

2. Describe the differences you observed in the type and distribution of hair on the front (anterior) and back (posterior) of your
   forearm. _____

   _____

3. Explain how a hair is formed. ⚠ _____

   _____

Part C Assessments

Complete the following:

1. Distinguish the locations and tissues among epidermis, dermis, and hypodermis layers. ⚠ _____

   _____

   _____

   _____

2. How do the cells of stratum corneum and stratum basale differ? ⚠ _____

   _____

   _____

3. State the specific location of melanin observed in dark skin. ⚠ _____

   _____

4. What special qualities does the connective tissue of the dermis have? ⚠ _____

   _____

Part D Assessments

Complete the following:

1. What part of the hair extends from the hair papilla to the body surface? ⚠ _____

2. In which layer of skin are sebaceous glands found? ⚠ _____

3. How are sebaceous glands associated with hair follicles? ⚠ _____

_____

4. In which layer of skin are sweat glands usually located? ⚠ _____

Part E Assessments

Sketch a vertical section of human skin, using the scanning objective. Label the skin layers and a hair follicle, a sebaceous gland, and a sweat gland. ⚠

Organization of the Skeleton

Purpose of the Exercise

To review the organization of the skeleton, the major bones of the skeleton, and the terms used to describe skeletal structures.

Materials Needed

Human skeleton, articulated
Human skeleton, disarticulated

For Learning Extension Activity:
Colored pencils

For Demonstration Activity:
Radiographs (X rays) of skeletal structures
Stereomicroscope (dissecting microscope)

Learning Outcomes

After completing this exercise, you should be able to

(1) Locate and label the major bones of the human skeleton.

(2) Distinguish between the axial skeleton and the appendicular skeleton.

(3) Associate the terms used to describe skeletal structures and locate examples of such structures on the human skeleton.

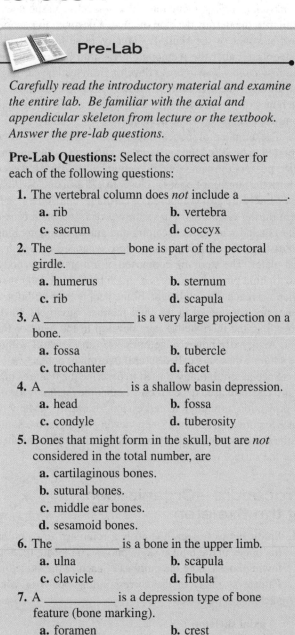

Pre-Lab

Carefully read the introductory material and examine the entire lab. Be familiar with the axial and appendicular skeleton from lecture or the textbook. Answer the pre-lab questions.

Pre-Lab Questions: Select the correct answer for each of the following questions:

1. The vertebral column does *not* include a _____.
   - **a.** rib
   - **b.** vertebra
   - **c.** sacrum
   - **d.** coccyx

2. The _____ bone is part of the pectoral girdle.
   - **a.** humerus
   - **b.** sternum
   - **c.** rib
   - **d.** scapula

3. A _____ is a very large projection on a bone.
   - **a.** fossa
   - **b.** tubercle
   - **c.** trochanter
   - **d.** facet

4. A _____ is a shallow basin depression.
   - **a.** head
   - **b.** fossa
   - **c.** condyle
   - **d.** tuberosity

5. Bones that might form in the skull, but are *not* considered in the total number, are
   - **a.** cartilaginous bones.
   - **b.** sutural bones.
   - **c.** middle ear bones.
   - **d.** sesamoid bones.

6. The _____ is a bone in the upper limb.
   - **a.** ulna
   - **b.** scapula
   - **c.** clavicle
   - **d.** fibula

7. A _____ is a depression type of bone feature (bone marking).
   - **a.** foramen
   - **b.** crest
   - **c.** sulcus
   - **d.** tuberosity

The skeleton can be divided into two major portions: (1) the *axial skeleton*, which consists of the bones and cartilages of the head, neck, and trunk, and (2) the *appendicular skeleton*, which consists of the bones of the limbs and those that anchor the limbs to the axial skeleton. The bones that anchor the limbs include the pectoral and pelvic girdles.

Ossification is the bone formation process. This process occurs during fetal development and continues through childhood growth. Bone forms either from intramembranous origins or endochondral origins. During the growth of *intramembranous bones*, membrane-like connective tissue layers similar to the dermis develop in an area destined to become the flat bones of the skull. Eventually, bone-forming cells (osteoblasts) alter the membrane areas into bone tissue. In contrast, *endochondral bones* develop from hyaline cartilage and through the ossification process develop into most of the bones of the skeleton. In either type of ossification, both compact and spongy bone develops.

The number of bones in the adult skeleton is often reported to be 206. Men and women, although variations can exist, possess the same total bone number of 206. However, at birth the number of bones is closer to 275 as many ossification centers are still composed of cartilage, and some bones form during childhood. For example, each hip bone (coxal bone) includes an ilium, ischium, and pubis, and many long bones have three ossification centers separated by epiphyseal plates. The sternum, composed of a manubrium, body, and xiphoid process, becomes a single bone much later than when we reach our full height. Some people have additional bones not considered in the total number. Sesamoid bones other than the the patellae may develop in the hand or the foot. Also, extra bones sometimes form in the skull within the sutures; these are called sutural (wormian) bones.

Special terminology is used to describe the features of a bone. The term used depends on whether the feature is a type of projection, articulation, depression, or opening. Many of these features can be noted when viewing radiographs. Some of the features can be palpated (touched) if they are located near the surface of the body.

Procedure—Organization of the Skeleton

1. Study figure 6.1 and use it as a reference to examine the bones in the human skeleton. As you locate the following bones, note the number of each in the skeleton. Palpate as many of the corresponding bones in your skeleton as possible.

   axial skeleton
   - skull
     - cranium (8)
     - face (14)
   - middle ear bone (6)
   - hyoid bone—supports the tongue (1)
   - vertebral column
     - vertebra (24)
     - sacrum (1)
     - coccyx (1)
   - thoracic cage
     - rib (24)
     - sternum (1)

   appendicular skeleton
   - pectoral girdle
     - scapula (2)
     - clavicle (2)
   - upper limbs
     - humerus (2)
     - radius (2)
     - ulna (2)
     - carpal (16)
     - metacarpal (10)
     - phalanx (28)
   - pelvic girdle
     - hip bone (coxal bone; pelvic bone; innominate bone) (2)
   - lower limbs
     - femur (2)
     - tibia (2)
     - fibula (2)
     - patella (2)
     - tarsal (14)
     - metatarsal (10)
     - phalanx (28)

Total	**206 bones**

2. Complete Part A of Laboratory Assessment 6.
3. Bone features (bone markings) can be grouped together in a category of projections, articulations, depressions, or openings. Within each category more specific examples occur. The bones illustrated in figure 6.2 represent specific examples of locations of specific features in the human body. Locate each of the following features on the example bone from a disarticulated skeleton, noting the size, shape, and location in the human skeleton.

   Projections: sites for tendon and ligament attachment
   crest—ridgelike
   epicondyle—superior to condyle
   line (linea)—slightly raised ridge
   process—prominent
   protuberance—outgrowth
   ramus—extension
   spine—thornlike
   trochanter—large

FIGURE 6.1 Major bones of the skeleton: (a) anterior view; (b) posterior view. The axial portion is orange, and the appendicular portion is yellow.

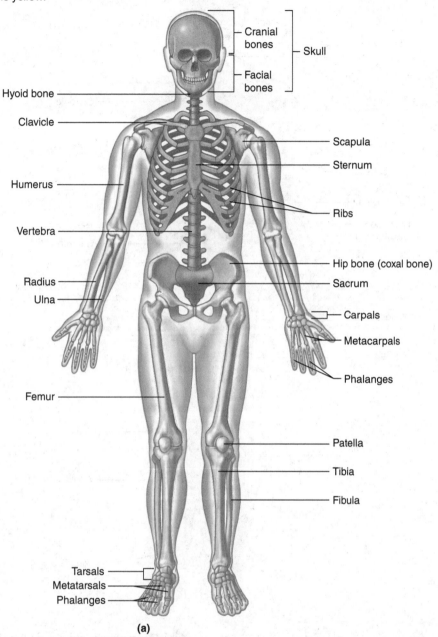

Cranial bones
Facial bones
Skull
Hyoid bone
Clavicle
Scapula
Sternum
Humerus
Ribs
Vertebra
Hip bone (coxal bone)
Radius
Sacrum
Ulna
Carpals
Metacarpals
Phalanges
Femur
Patella
Tibia
Fibula
Tarsals
Metatarsals
Phalanges

(a)

tubercle—small knoblike
tuberosity—rough elevation

<u>Articulations:</u> **where bones connect at a joint or articulate with each other**
condyle—rounded process
facet—nearly flat
head—expanded end

<u>Depressions:</u> **recessed areas in bones**
alveolus—socket
fossa—shallow basin

fovea—tiny pit
notch—indentation on edge
sulcus—narrow groove

<u>Openings:</u> **open spaces in bones**
canal—tubular passage
fissure—slit
foramen—hole
meatus—tubelike opening
sinus—cavity

FIGURE 6.1 *Continued.*

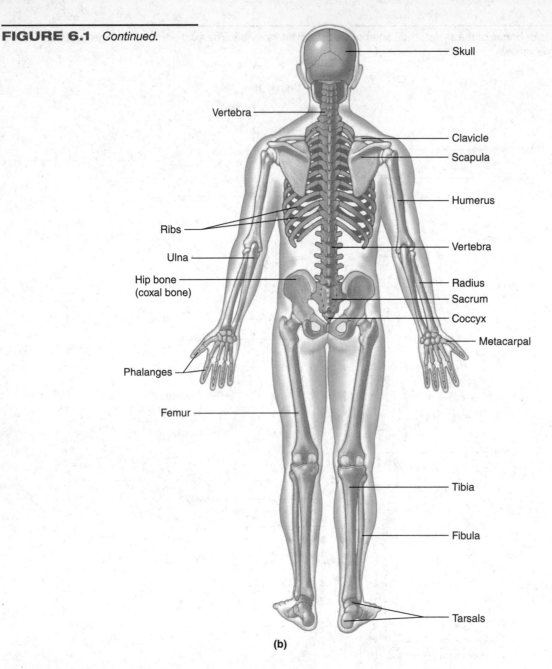

Skull

Vertebra

Clavicle

Scapula

Humerus

Ribs

Vertebra

Ulna

Hip bone
(coxal bone)

Radius

Sacrum

Coccyx

Metacarpal

Phalanges

Femur

Tibia

Fibula

Tarsals

(b)

 Critical Thinking Activity

Locate and name the largest foramen in the skull.

_____

_____

Locate and name the largest foramen in the skeleton.

_____

_____

4. Complete Parts B, C, and D of the laboratory assessment.

Demonstration Activity

Images on radiographs (X rays) are produced by allowing X rays from an X-ray tube to pass through a body part and to expose photographic film positioned on the opposite side of the part. The image that appears on the film after it is developed reveals the presence of parts with different densities. Bone, for example, is very dense tissue and is a good absorber of X rays. Thus, bone generally appears light on the film. Air-filled spaces, on the other hand, absorb almost no X rays and appear as dark areas on the film. Liquids and soft tissues absorb intermediate quantities of X rays, so they usually appear in various shades of gray.

Examine the available radiographs of skeletal structures by holding each film in front of a light source. Identify as many of the bones and features as you can.

FIGURE 6.2 Representative examples of bone features (bone markings) on bones of the skeleton (a–h). The two largest foramina of the skeleton have complete labels.

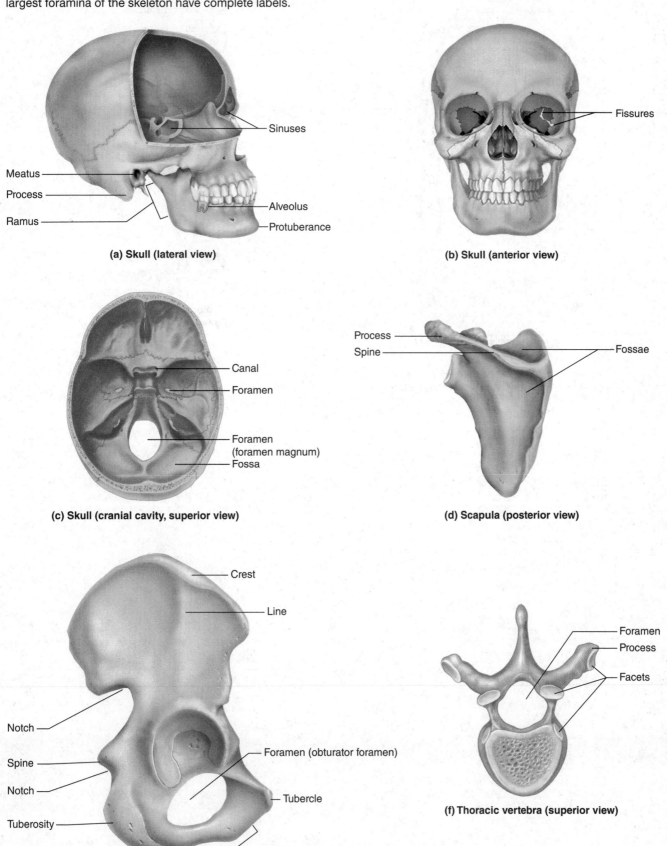

Sinuses

Meatus
Process
Ramus
Alveolus
Protuberance

(a) Skull (lateral view)

Fissures

(b) Skull (anterior view)

Canal
Foramen

Foramen (foramen magnum)
Fossa

(c) Skull (cranial cavity, superior view)

Process
Spine
Fossae

(d) Scapula (posterior view)

Crest
Line

Notch
Spine
Notch

Foramen (obturator foramen)

Tubercle

Tuberosity

Ramus

(e) Hip bone (lateral view)

Foramen
Process
Facets

(f) Thoracic vertebra (superior view)

FIGURE 6.2 *Continued.*

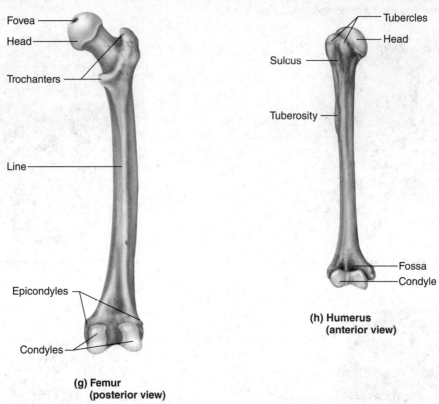

Fovea

Head

Trochanters

Line

Epicondyles

Condyles

(g) Femur
(posterior view)

Tubercles

Head

Sulcus

Tuberosity

Fossa

Condyle

(h) Humerus
(anterior view)

Name _____

Date _____

Section _____

The ⚠ corresponds to the indicated outcome(s) found at the beginning of the laboratory exercise.

Organization of the Skeleton

Part A   Assessments

Label the bones indicated in figure 6.3. ⚠

FIGURE 6.3   Label the major bones of the skeleton: (a) anterior view; (b) posterior view, using the terms provided. ⚠ ⚠

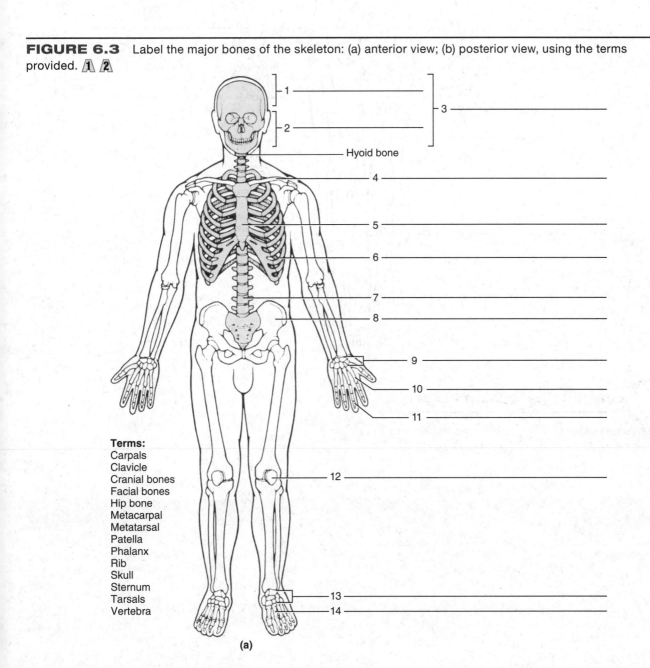

1 _____

2 _____

3 _____

Hyoid bone

4 _____

5 _____

6 _____

7 _____

8 _____

9 _____

10 _____

11 _____

12 _____

13 _____

14 _____

Terms:
Carpals
Clavicle
Cranial bones
Facial bones
Hip bone
Metacarpal
Metatarsal
Patella
Phalanx
Rib
Skull
Sternum
Tarsals
Vertebra

(a)

FIGURE 6.3 *Continued.*

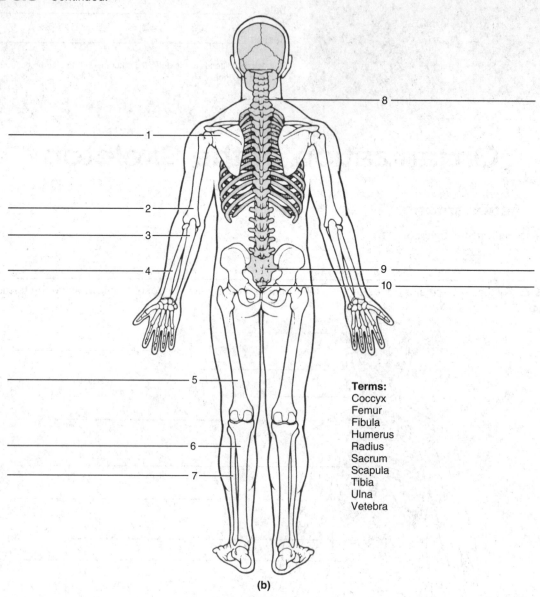

1 _____

2 _____

3 _____

4 _____

5 _____

6 _____

7 _____

8 _____

9 _____

10 _____

Terms:
Coccyx
Femur
Fibula
Humerus
Radius
Sacrum
Scapula
Tibia
Ulna
Vetebra

(b)

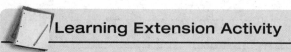

Learning Extension Activity

Use colored pencils to distinguish the individual bones in figure 6.3.

Part B Assessments

1. Identify the bones indicated in figure 6.4.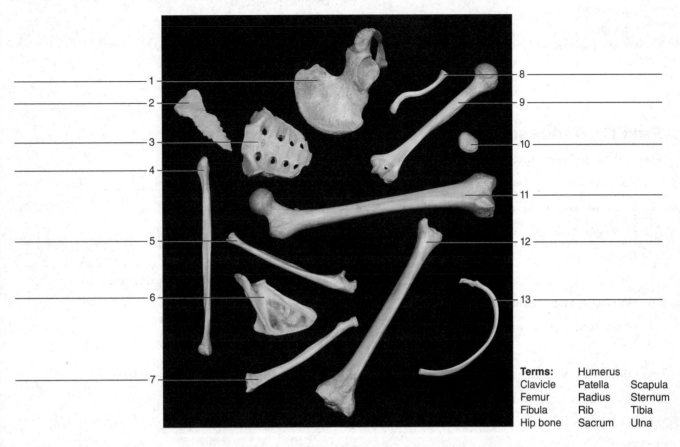

FIGURE 6.4   Identify the bones in this random arrangement, using the terms provided.

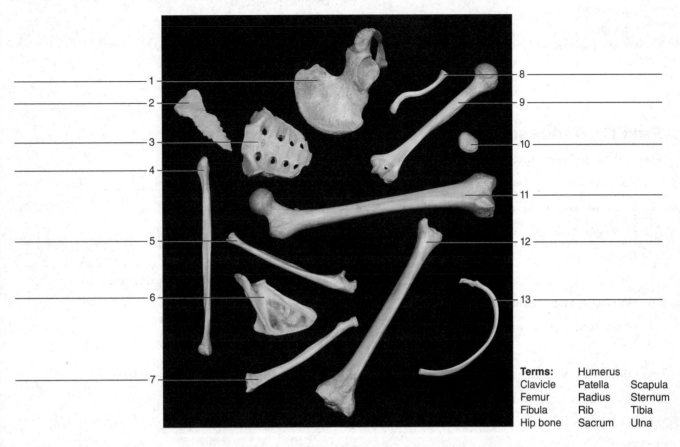

1 _____
2 _____
3 _____
4 _____
5 _____
6 _____
7 _____
8 _____
9 _____
10 _____
11 _____
12 _____
13 _____

Terms: Humerus
Clavicle   Patella   Scapula
Femur   Radius   Sternum
Fibula   Rib   Tibia
Hip bone   Sacrum   Ulna

2. List any of the bones shown in figure 6.4 that are included as part of the axial skeleton. _____

_____

Part C Assessments

1. Match the terms in column A with the definitions in column B. Place the letter of your choice in the space provided.

Column A	Column B
a. Condyle	_____ **1.** Small, nearly flat articular surface
b. Crest	_____ **2.** Deep depression or shallow basin
c. Facet	_____ **3.** Rounded process
d. Foramen	_____ **4.** Opening or hole
e. Fossa	_____ **5.** Projection extension
f. Line	_____ **6.** Ridgelike projection
g. Ramus	_____ **7.** Slightly raised ridge

55

2. Match the terms in column A with the definitions in column B. Place the letter of your choice in the space provided. 🔺

Column A	Column B
a. Fovea	_____ **1.** Tubelike opening
b. Head	_____ **2.** Tiny pit or depression
c. Meatus	
d. Sinus	_____ **3.** Small, knoblike projection
e. Spine	_____ **4.** Thornlike projection
f. Trochanter	_____ **5.** Rounded enlargement at end of bone
g. Tubercle	_____ **6.** Air-filled cavity within bone
	_____ **7.** Relatively large process

Part D Assessments

Complete the following statements:

1. The extra bones that sometimes develop between the flat bones of the skull are called _____ bones. 🔺

2. Small bones occurring in some tendons are called _____ bones. 🔺

3. The cranium and facial bones compose the _____. 🔺

4. The _____ bone supports the tongue. 🔺

5. The _____ at the inferior end of the sacrum is composed of several fused vertebrae. 🔺

6. Most ribs are attached anteriorly to the _____. 🔺

7. The thoracic cage is composed of _____ pairs of ribs. 🔺

8. The scapulae and clavicles together form the _____. 🔺

9. Which of the following bones is *not* part of the appendicular skeleton: clavicle, femur, scapula, sternum? _____ 🔺

10. The wrist is composed of eight bones called _____. 🔺

11. The hip bones (coxal bones) are attached posteriorly to the _____. 🔺

12. The _____ bone covers the anterior surface of the knee. 🔺

13. The bones that articulate with the distal ends of the tibia and fibula are called _____. 🔺

14. All finger and toe bones are called _____. 🔺

Vertebral Column and Thoracic Cage

Purpose of the Exercise

To examine the vertebral column and the thoracic cage of the human skeleton and to identify the bones and major features of these parts.

Materials Needed

Human skeleton, articulated
Samples of cervical, thoracic, and lumbar vertebrae
Human skeleton, disarticulated

Learning Outcomes

After completing this exercise, you should be able to

1. Identify the structures and functions of the vertebral column.

2. Locate the features of a vertebra.

3. Distinguish the cervical, thoracic, and lumbar vertebrae and the sacrum and coccyx.

4. Identify the structures and functions of the thoracic cage.

5. Distinguish between true and false ribs.

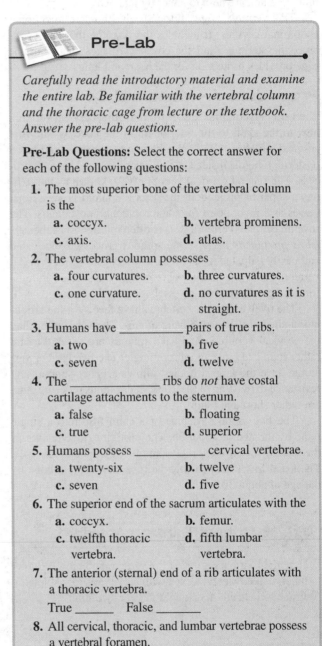

Pre-Lab

Carefully read the introductory material and examine the entire lab. Be familiar with the vertebral column and the thoracic cage from lecture or the textbook. Answer the pre-lab questions.

Pre-Lab Questions: Select the correct answer for each of the following questions:

1. The most superior bone of the vertebral column is the
   a. coccyx.   b. vertebra prominens.
   c. axis.   d. atlas.

2. The vertebral column possesses
   a. four curvatures.   b. three curvatures.
   c. one curvature.   d. no curvatures as it is straight.

3. Humans have _____ pairs of true ribs.
   a. two   b. five
   c. seven   d. twelve

4. The _____ ribs do *not* have costal cartilage attachments to the sternum.
   a. false   b. floating
   c. true   d. superior

5. Humans possess _____ cervical vertebrae.
   a. twenty-six   b. twelve
   c. seven   d. five

6. The superior end of the sacrum articulates with the
   a. coccyx.   b. femur.
   c. twelfth thoracic   d. fifth lumbar
      vertebra.   vertebra.

7. The anterior (sternal) end of a rib articulates with a thoracic vertebra.
   True _____ False _____

8. All cervical, thoracic, and lumbar vertebrae possess a vertebral foramen.
   True _____ False _____

9. A feature of the second cervical vertebra is the dens.
   True _____ False _____

The vertebral column, consisting of twenty-six bones, extends from the skull to the pelvis and forms the vertical axis of the human skeleton. The *vertebral column* includes seven cervical vertebrae, twelve thoracic vertebrae, five lumbar vertebrae, one sacrum of five fused vertebrae, and one coccyx of usually four fused vertebrae. To help you to remember the number of cervical, thoracic, and lumbar vertebrae from superior to inferior, consider this saying: breakfast at 7, lunch at 12, and dinner at 5. These vertebrae are separated from one another by cartilaginous intervertebral discs and are held together by ligaments.

The *thoracic cage* surrounds the thoracic and upper abdominal cavities. It includes the ribs, the thoracic vertebrae, the sternum, and the costal cartilages. The thoracic cage provides protection for the heart and lungs.

Procedure A—Vertebral Column

The vertebral column extends from the first cervical vertebra next to the skull to the inferior tip of the coccyx. The first cervical vertebra (C1) is also known as the *atlas* and has a posterior tubercle instead of a more pronounced spinous process. The second cervical vertebra (C2), known as the *axis,* has a superior projection, the dens (odontoid process) that serves as a pivot point for some rotational movements. The seventh cervical vertebra (C7) is often referred to as the *vertebra prominens* because the spinous process is elongated and easily palpated as a surface feature. The seven cervical vertebrae have the distinctive feature of transverse foramina for passageways of blood vessels serving the brain.

The twelve thoracic vertebrae have facets for the articulation sites of the twelve pairs of ribs. They are larger than the cervical vertebrae and have spinous processes that are rather long and have an inferior angle. The five lumbar vertebrae have the largest bodies, allowing better support and resistance to twisting of the trunk, and spinous processes that are rather short and blunt.

The five sacral vertebrae of a child fuse into a single bone by the age of about 26. The posterior ridge known as the *medial sacral crest* represents fused spinous processes. The usual four coccyx vertebrae fuse into a single bone by the age of about 30.

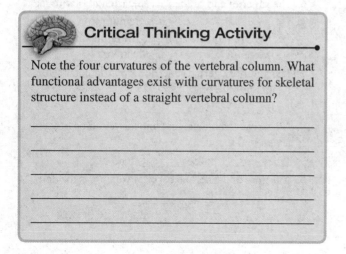

Critical Thinking Activity

Note the four curvatures of the vertebral column. What functional advantages exist with curvatures for skeletal structure instead of a straight vertebral column?

_____

_____

_____

_____

_____

The four curvatures of the vertebral column develop either before or after birth. The thoracic and sacral curvatures (primary curvatures) form by the time of birth. The cervical curvature develops by the time a baby is able to hold the head erect and crawl, while the lumbar curvature forms by the time the child is able to walk. The cervical and lumbar curvatures represent the secondary curvatures. The four curvatures allow for flexibility and resiliency of the vertebral column and for it to function somewhat like a spring instead of a rigid rod.

1. Examine figure 7.1 and the vertebral column of the human skeleton. Locate the following bones and features. At the same time, locate as many of the cor-

FIGURE 7.1 Bones and features of the vertebral column (right lateral view).

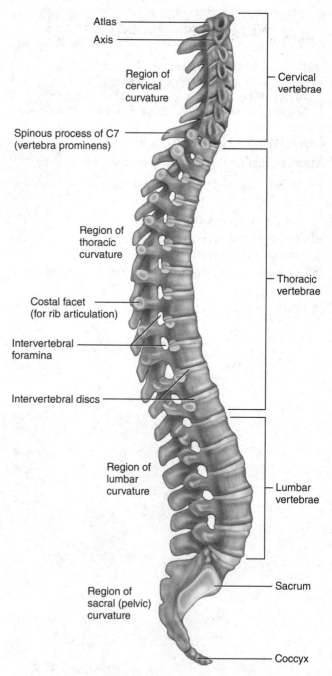

Atlas
Axis
Region of cervical curvature
Cervical vertebrae
Spinous process of C7 (vertebra prominens)
Region of thoracic curvature
Thoracic vertebrae
Costal facet (for rib articulation)
Intervertebral foramina
Intervertebral discs
Region of lumbar curvature
Lumbar vertebrae
Region of sacral (pelvic) curvature
Sacrum
Coccyx

responding bones and features in your skeleton as possible.

cervical vertebrae **(7)**
- atlas (C1)
- axis (C2)
- vertebra prominens (C7)

thoracic vertebrae **(12)**

lumbar vertebrae **(5)**

sacrum **(1)**

coccyx **(1)**

intervertebral discs—fibrocartilage

vertebral canal—contains spinal cord

cervical curvature

thoracic curvature

lumbar curvature

sacral (pelvic) curvature

intervertebral foramina—passageway for spinal nerves

2. Compare the available samples of cervical, thoracic, and lumbar vertebrae along with figures 7.2 and 7.3. Note differences in size and shapes and locate the following features:

vertebral foramen—location of spinal cord

body—largest part of vertebra; main support portion

pedicles—form lateral area of vertebral foramen

laminae—thin plates forming posterior area of vertebral foramen

spinous process—posterior projection

transverse processes—lateral projections

facets—articulating surfaces

superior articular processes—superior projections

inferior articular processes—inferior projections

inferior vertebral notch—space for nerve passage

transverse foramina—only present on cervical vertebrae; passageway for blood vessels

dens (odontoid process) of axis—superior process of C2 and is a pivot location at atlas

---

FIGURE 7.2 The superior features of (a) the atlas, and the superior and right lateral features of (b) the axis. (The broken arrow indicates a transverse foramen.)

Number code:
1. Superior articular facet
2. Vertebral foramen
3. Dens (odontoid process)
4. Facet that articulates with occipital condyle
5. Transverse process
6. Transverse foramen
7. Spinous process
8. Body

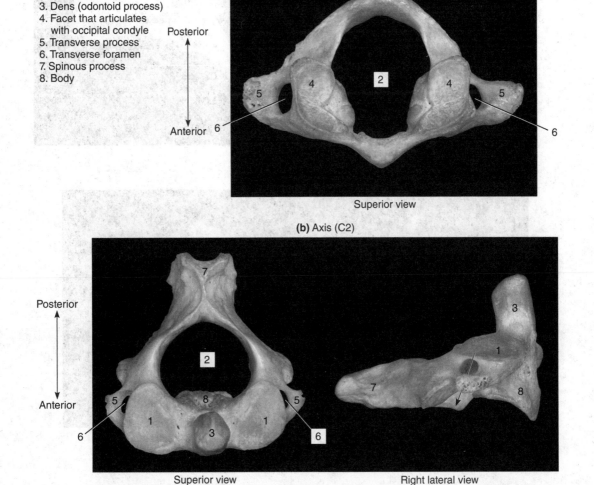

(a) Atlas (C1)

Superior view

(b) Axis (C2)

Superior view

Right lateral view

Posterior ◄──────► Anterior

FIGURE 7.3 The superior and right lateral features of the (a) cervical, (b) thoracic, and (c) lumbar vertebrae. (The broken arrow indicates a transverse foramen.)

Number code:
1. Superior articular process
2. Transverse process
3. Lamina
4. Spinous process
5. Pedicle
6. Body
7. Inferior vertebral notch
8. Vertebral foramen
9. Transverse foramen
10. Costal facets

Superior views   Right lateral views

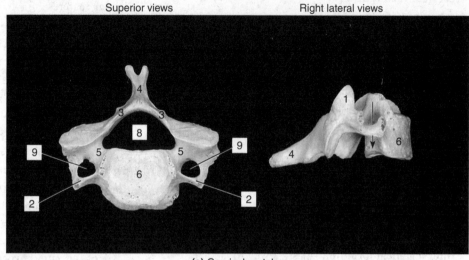

(a) Cervical vertebra

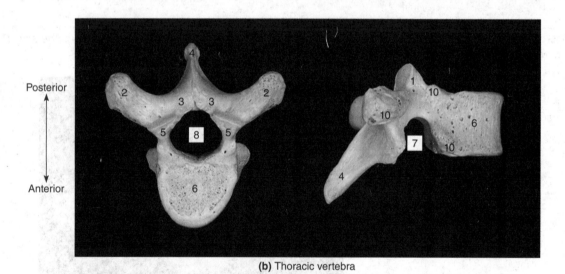

(b) Thoracic vertebra

(c) Lumbar vertebra

FIGURE 7.4 The sacrum and coccyx.

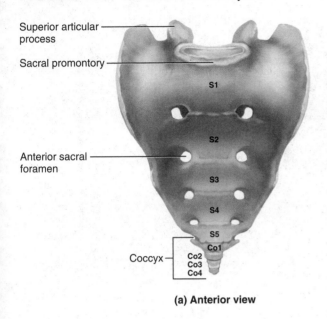

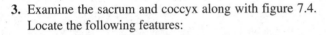

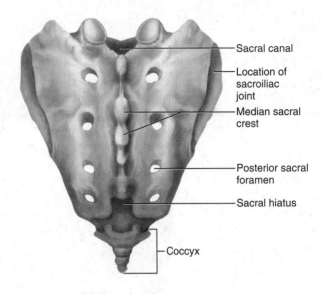

(a) Anterior view

(b) Posterior view

3. Examine the sacrum and coccyx along with figure 7.4. Locate the following features:

sacrum—formed by five fused vertebrae
- superior articular process—superior projection with a facet articulation site
- anterior sacral foramen—passageway for blood vessels and nerves
- posterior sacral foramen—passageway for blood vessels and nerves
- sacral promontory—anterior border of S1; important landmark for obstetricians
- sacral canal—portion of vetrebral canal
- median sacral crest—area of fused spinous processes
- sacral hiatus—inferior opening of vertebral canal

coccyx—formed by three to five fused vertebrae

4. Complete Parts A and B of Laboratory Assessment 7.

Procedure B—Thoracic Cage

The twelve thoracic vertebrae are associated with the twelve pairs of ribs. The superior seven pairs of ribs are connected directly to the sternum with costal cartilage and are known as true ribs. The five inferior pairs, known as false ribs, either connect indirectly to the sternum with costal cartilage or do not connect to the sternum. Pairs eleven and twelve are called the floating ribs because they only connect with the thoracic vertebrae and not with the sternum.

The manubrium, body, and xiphoid process represent three regions that eventually fuse into a single flat bone, the sternum. The heart is located mainly beneath the body portion of the sternum. Any chest compressions adminis-

tered during cardiopulmonary resuscitation should be over the body area of the sternum, not the xiphoid process region. Chest compressions over the xiphoid process region can force the xiphoid process deep into the liver or the inferior portion of the heart and cause a fatal hemorrhage to occur.

1. Examine figures 7.5 and 7.6 and the thoracic cage of the human skeleton. Locate the following bones and features:

rib
- head—expanded end near thoracic vertebra
- tubercle—projection near thoracic vertebra
- neck—narrow region between head and tubercle
- shaft—main portion
- anterior (sternal) end—costal cartilage location
- facets—articulation surfaces
- true ribs—pairs 1–7
- false ribs—pairs 8–12; includes floating ribs
  - floating ribs—pairs 11–12

costal cartilages—hyaline cartilage

sternum
- jugular (suprasternal) notch—superior concave border of manubrium
- clavicular notch—articulation site of clavicle
- manubrium—superior part
- sternal angle—junction of manubrium and body at level of second rib pair
- body—largest, middle part
- xiphoid process—inferior part; remains cartilaginous until adulthood

2. Complete Parts C and D of the laboratory assessment.

FIGURE 7.5 Superior view of a typical rib with the articulation sites with a thoracic vertebra.

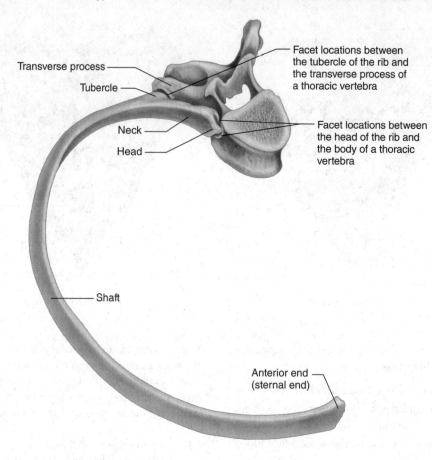

Transverse process

Tubercle

Neck

Head

Facet locations between the tubercle of the rib and the transverse process of a thoracic vertebra

Facet locations between the head of the rib and the body of a thoracic vertebra

Shaft

Anterior end (sternal end)

FIGURE 7.6 Bones and features of the thoracic cage (anterior view).

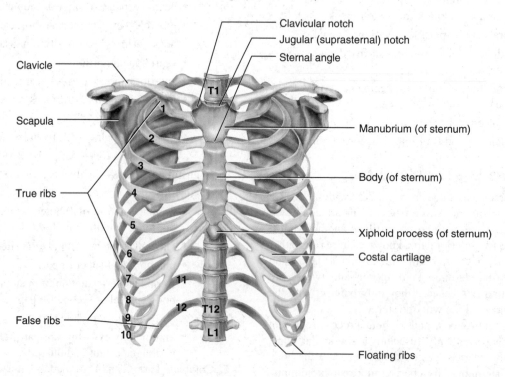

Clavicular notch

Jugular (suprasternal) notch

Sternal angle

Clavicle

Scapula

True ribs

False ribs

T1

Manubrium (of sternum)

Body (of sternum)

Xiphoid process (of sternum)

Costal cartilage

T12

L1

Floating ribs

Name _____

Date _____

Section _____

The ⚠ corresponds to the indicated outcome(s) found at the beginning of the laboratory exercise.

Vertebral Column and Thoracic Cage

Part A Assessments

Complete the following statements:

1. The vertebral column encloses and protects the _____. ⚠

2. The vertebral column extends from the skull to the _____. ⚠

3. The seventh cervical vertebra is called the _____ and has an obvious spinous process surface feature that can be palpated. ⚠

4. The _____ of the vertebrae support the weight of the head and trunk. ⚠

5. The _____ separate adjacent vertebrae, and they soften the forces created by walking. ⚠

6. The intervertebral foramina provide passageways for _____. ⚠

7. Transverse foramina of _____ vertebrae serve as passageways for blood vessels leading to the brain. ⚠

8. The first vertebra also is called the _____. ⚠

9. When the head is moved from side to side, the first vertebra pivots around the _____ of the second vertebra. ⚠

10. The _____ vertebrae have the largest and strongest bodies. ⚠

11. The number of vertebrae that fuse in the adult to form the sacrum is _____. ⚠

Part B Assessments

1. Based on your observations, compare typical cervical, thoracic, and lumbar vertebrae in relation to the characteristics indicated in the table. The table is partly completed. For your responses, consider characteristics such as size, shape, presence or absence, and unique features. ⚠ ⚠

Vertebra	Number	Size	Body	Spinous Process	Transverse Foramina
Cervical	7		smallest	C2 through C6 are forked	
Thoracic		intermediate			
Lumbar					absent

2. Identify the bones and features in figures 7.7 and 7.8.

FIGURE 7.7 Label the bones and features of a lateral view of a vertebral column by placing the correct numbers in the spaces provided. 🛆 🛆

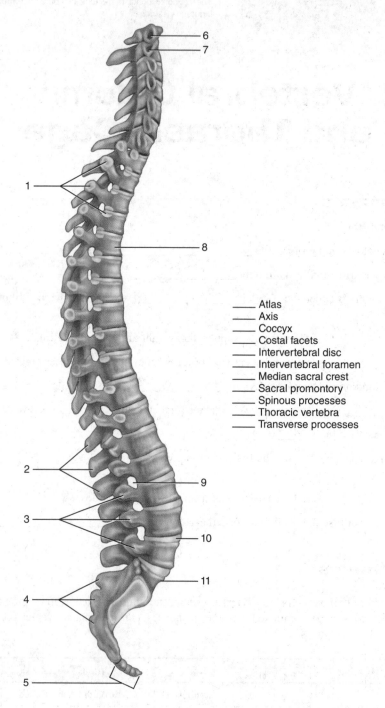

_____ Atlas
_____ Axis
_____ Coccyx
_____ Costal facets
_____ Intervertebral disc
_____ Intervertebral foramen
_____ Median sacral crest
_____ Sacral promontory
_____ Spinous processes
_____ Thoracic vertebra
_____ Transverse processes

FIGURE 7.8 Identify the bones and features indicated in this radiograph of the neck (lateral view), using the terms provided. ⚠️ ⚠️

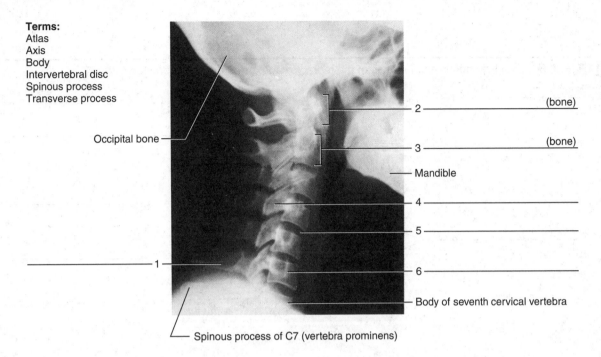

Terms:
Atlas
Axis
Body
Intervertebral disc
Spinous process
Transverse process

Occipital bone ———

2 _____ (bone)

3 _____ (bone)

— Mandible

4 _____

5 _____

1 ———

6 _____

— Body of seventh cervical vertebra

— Spinous process of C7 (vertebra prominens)

Part C   Assessments

Complete the following statements:

1. The manubrium, body, and xiphoid process form a bone called the _____. 🅐

2. The last two pairs of ribs that have no cartilaginous attachments to the sternum are sometimes called _____ ribs. 🄂

3. There are _____ pairs of true ribs. 🄂

4. Costal cartilages are composed of _____ tissue. 🅐

5. The manubrium articulates with the _____ on its superior border. 🅐

6. List three general functions of the thoracic cage. 🅐 _____

_____

7. The sternal angle indicates the location of the _____ pair of ribs. 🅐

Part D Assessments

Identify the bones and features indicated in figure 7.9.

FIGURE 7.9 Label the bones and features of the thoracic cage, using the terms provided. ⒶⒺ

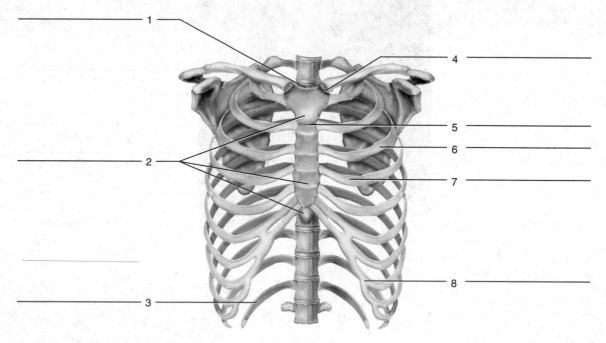

Terms:
Costal cartilage of false rib
Costal cartilage of true rib
Clavicular notch
Floating rib
Jugular notch
Sternal angle
Sternum
True rib

Pectoral Girdle and Upper Limb

Purpose of the Exercise

To examine the bones of the pectoral girdle and upper limb and to identify the major features of these bones.

Materials Needed

Human skeleton, articulated
Human skeleton, disarticulated

Learning Outcomes

After completing this exercise, you should be able to

① Locate and identify the bones of the pectoral girdle and their major features.

② Locate and identify the bones of the upper limb and their major features.

Pre-Lab

Carefully read the introductory material and examine the entire lab. Be familiar with the pectoral girdle and upper limb bones from lecture or the textbook. Answer the pre-lab questions.

Pre-Lab Questions: Select the correct answer for each of the following questions:

1. The clavicle and the scapula form the
   a. pectoral girdle.   b. pelvic girdle.
   c. upper limb.   d. axial skeleton.

2. Anatomically, *arm* represents
   a. shoulder to fingers. b. shoulder to wrist.
   c. elbow to wrist.   d. shoulder to elbow.

3. Which of the following is *not* part of the scapula?
   a. spine   b. manubrium
   c. acromion   d. supraspinous fossa

4. Which carpal is included in the proximal row?
   a. hamate   b. capitate
   c. trapezium   d. lunate

5. Which of the following is the most proximal part of the upper limb?
   a. styloid process of radius
   b. styloid process of ulna
   c. head of humerus
   d. medial epicondyle of humerus

6. Which of the following is the most distal feature of the humerus?
   a. anatomical neck   b. deltoid tuberosity
   c. capitulum   d. head

7. The capitate is one of the eight carpals in a wrist.
   True _____   False _____

8. The clavicle articulates with the sternum and the humerus.
   True _____   False _____

A **pectoral girdle (shoulder girdle)** consists of an anterior clavicle and a posterior scapula. A *pectoral girdle* represents an incomplete ring (girdle) of bones as the posterior scapulae do not meet each other, but muscles extend from their medial borders to the vertebral column. The clavicles on their medial ends form a joint with the manubrium of the sternum. The pectoral girdle supports the upper limb and serves as attachments for various muscles that move the upper limb. This allows considerable flexibility of the shoulder. Relatively loose attachments of the pectoral girdle with the humerus allow a wide range of movements, but shoulder joint injuries are somewhat common. Additionally, the clavicle is a frequently broken bone when one reaches with an upper limb to break a fall.

Each upper limb includes a humerus in the arm, a radius and ulna in the forearm, and eight carpals, five metacarpals, and fourteen phalanges in the hand. (Anatomically, *arm* represents the region from shoulder to elbow, *forearm* is elbow to the wrist, and *hand* includes the wrist to the ends of digits.) These bones form the framework of the upper limb. They also function as parts of levers when muscles contract.

Procedure A—Pectoral Girdle

The clavicle and scapula of the pectoral girdle provide for attachments of neck and trunk muscles. The clavicle is not a straight bone, but rather has two curves making it slightly S-shaped. The clavicle serves as a brace bone to keep the upper limb to the side of the body. The bone is easily fractured because it is so close to the anterior surface and because the shoulder or upper limb commonly breaks a fall. Fortunately bones have tensile strength, and when a force is exerted upon the clavicle from the lateral side of the body, the bone will bend to some extent at both curves until the threshold is reached, causing a fracture to occur.

The scapula has many tendon attachment sites for muscles of the neck, trunk, and upper limb. Because the scapula does not connect directly to the axial skeleton and because it has so many muscle attachments, there is a great deal of flexibility for shoulder movements. Unfortunately, this much flexibility can result in dislocating the humerus from the scapula. This type of injury can occur during gymnastics maneuvers or when suddenly pulling or swinging a child by an upper limb.

1. Examine figures 8.1 and 8.2 along with the bones of the pectoral girdle. Locate the following features of the clavicle and the scapula. At the same time, locate as many of the corresponding surface bones and features of your own skeleton as possible.

   clavicle
   - sternal (medial) end—articulates with the manubrium of the sternum
   - acromial (lateral) end—articulates with the acromion of the scapula

---

FIGURE 8.1   Bones and features of the right shoulder and upper limb (anterior view).

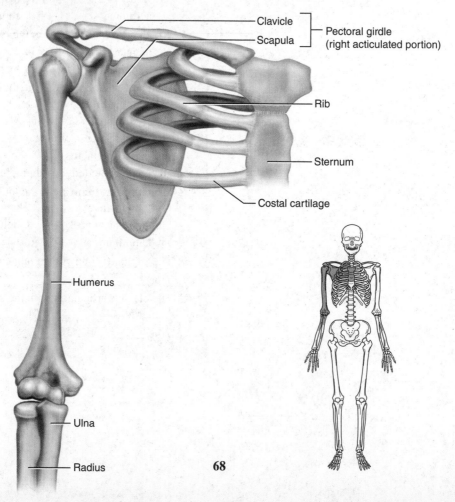

68

scapula
- spine
- acromion—lateral end of spine
- glenoid cavity—shallow socket; articulates with head of humerus
- coracoid process—beaklike projection
- borders
    - superior border
    - medial (vertebral) border
    - lateral (axillary) border
- fossae—shallow depressions
    - supraspinous fossa
    - infraspinous fossa
    - subscapular fossa
- angles
    - superior angle
    - inferior angle
- tubercles
    - supraglenoid tubercle
    - infraglenoid tubercle

2. Complete Parts A and B of Laboratory Assessment 8.

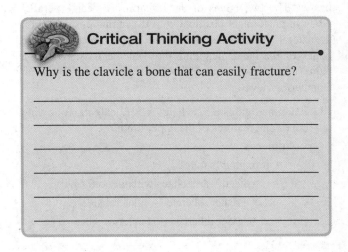

Critical Thinking Activity

Why is the clavicle a bone that can easily fracture?

_____

_____

_____

_____

_____

_____

FIGURE 8.2 Right scapula (a) posterior surface (b) lateral view (c) anterior surface.

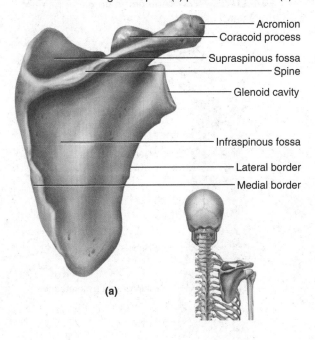

- Acromion
- Coracoid process
- Supraspinous fossa
- Spine
- Glenoid cavity
- Infraspinous fossa
- Lateral border
- Medial border

(a)

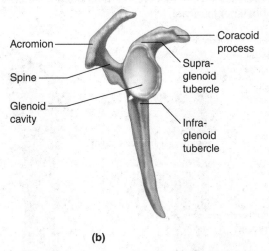

- Acromion
- Spine
- Glenoid cavity
- Coracoid process
- Supra-glenoid tubercle
- Infra-glenoid tubercle

(b)

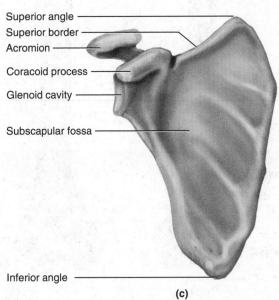

- Superior angle
- Superior border
- Acromion
- Coracoid process
- Glenoid cavity
- Subscapular fossa
- Inferior angle

(c)

Procedure B—Upper Limb

The humerus has sometimes been called the "funny bone" because of a tingling sensation (temporary pain) if it is bumped on the medial epicondyle where the ulnar nerve passes. The two bones of the forearm are nearly parallel when in anatomical position, with the radius positioned on the lateral side and the ulna on the medial side. All of the bones of the hand, including those of the wrist (carpals), palm (metacarpals), and digits (phalanges), are visible from an anterior view.

1. Examine figures 8.3, 8.4, and 8.5. Locate the following bones and features of the upper limb:

    humerus
    - proximal features
        - head—articulates with glenoid cavity
        - greater tubercle—on lateral side
        - lesser tubercle—on anterior side

- anatomical neck—tapered region near head
- surgical neck—common fracture site
- intertubercular sulcus—furrow for tendon of biceps muscle
- shaft
    - deltoid tuberosity
- distal features
    - capitulum—lateral condyle; articulates with radius
    - trochlea—medial condyle; articulates with ulna
    - medial epicondyle
    - lateral epicondyle
    - coronoid fossa—articulates with coronoid process of ulna
    - olecranon fossa—articulates with olecranon process of ulna

---

FIGURE 8.3 Right humerus (a) anterior features and (b) posterior features.

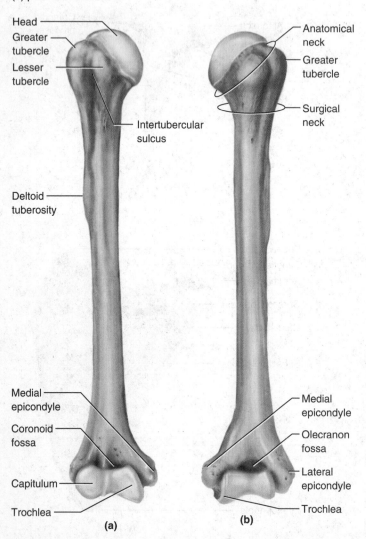

Head
Greater tubercle
Lesser tubercle
Intertubercular sulcus
Anatomical neck
Greater tubercle
Surgical neck
Deltoid tuberosity
Medial epicondyle
Coronoid fossa
Capitulum
Trochlea
Medial epicondyle
Olecranon fossa
Lateral epicondyle
Trochlea

(a)   (b)

FIGURE 8.4 Anterior features of the right radius and ulna.

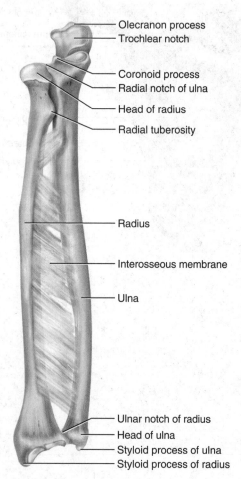

Olecranon process
Trochlear notch
Coronoid process
Radial notch of ulna
Head of radius
Radial tuberosity
Radius
Interosseous membrane
Ulna
Ulnar notch of radius
Head of ulna
Styloid process of ulna
Styloid process of radius

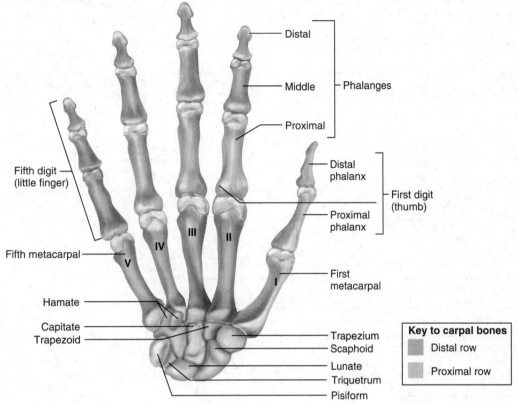

radius—lateral bone of forearm
- head of radius—allows rotation at elbow
- radial tuberosity
- styloid process of radius
- ulnar notch of radius—articulation site with ulna

ulna—medial bone of forearm; longer than radius
- trochlear notch
- radial notch of ulna—articulation site with head of radius
- olecranon process
- coronoid process
- styloid process of ulna
- head of ulna—at distal end

carpal bones—positioned in two irregular rows
- proximal row (listed lateral to medial)
  - scaphoid
  - lunate
  - triquetrum
  - pisiform
- distal row (listed medial to lateral)
  - hamate
  - capitate
  - trapezoid
  - trapezium

The following mnemonic device will help you learn the eight carpals:

So Long Top Part
Here Comes The Thumb

The first letter of each word corresponds to the first letter of a carpal. This device arranges the carpals in order for the proximal, transverse row of four bones from lateral to medial, followed by the distal, transverse row from medial to lateral, which ends nearest the thumb. This arrangement assumes the hand is in the anatomical position.

metacarpals (I–V)

phalanges—located in digits (fingers)
- proximal phalanx
- middle phalanx—not present in first digit
- distal phalanx
2. Complete Parts C, D, and E of the laboratory assessment.

NOTES

Name _____

Date _____

Section _____

The Ⓐ corresponds to the indicated outcome(s) found at the beginning of the laboratory exercise.

Pectoral Girdle and Upper Limb

Part A   Assessments

Complete the following statements:

1. The pectoral girdle is an incomplete ring because it is open in the back between the _____. Ⓐ

2. The medial end of a clavicle articulates with the _____ of the sternum. Ⓐ

3. The lateral end of a clavicle articulates with the _____ process of the scapula. Ⓐ

4. The _____ is a bone that serves as a brace between the sternum and the scapula. Ⓐ

5. The _____ divides the scapula into unequal portions. Ⓐ

6. The lateral tip of the shoulder is the _____ of the scapula. Ⓐ

7. Near the lateral end of the scapula, the _____ process of the scapula curves anteriorly and inferiorly from the clavicle. Ⓐ

8. The glenoid cavity of the scapula articulates with the _____ of the humerus. Ⓐ

Part B   Assessments

Label the structures indicated in figure 8.6.

FIGURE 8.6   Label the posterior surface of the right scapula, using the terms provided. Ⓐ

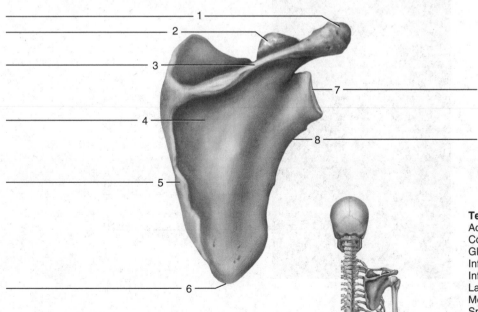

Terms:
Acromion
Coracoid process
Glenoid cavity
Inferior angle
Infraspinous fossa
Lateral border
Medial border
Spine

Part C Assessments

Match the bones in column A with the bones and features in column B. Place the letter of your choice in the space provided. 🄰

Column A	Column B
a. Carpals	_____ **1.** Capitate
b. Humerus	_____ **2.** Coronoid fossa
c. Metacarpals	_____ **3.** Deltoid tuberosity
d. Phalanges	_____ **4.** Greater tubercle
e. Radius	_____ **5.** Five palmar bones
f. Ulna	_____ **6.** Fourteen bones in digits
	_____ **7.** Intertubercular sulcus
	_____ **8.** Lunate
	_____ **9.** Olecranon fossa
	_____ **10.** Radial tuberosity
	_____ **11.** Trapezium
	_____ **12.** Trochlear notch

Part D Assessments

Identify the bones and features indicated in the radiographs of figures 8.7, 8.8, and 8.9.

FIGURE 8.7 Identify the bones and features indicated on this radiograph of the right elbow (anterior view), using the terms provided. 🄰

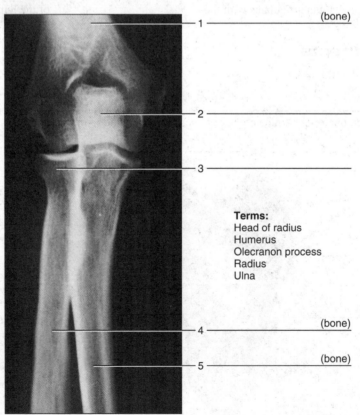

1 _____ (bone)

2 _____

3 _____

Terms:
Head of radius
Humerus
Olecranon process
Radius
Ulna

4 _____ (bone)

5 _____ (bone)

FIGURE 8.8 Identify the bones and features indicated on this radiograph of the anterior view of the right shoulder, using the terms provided. ⚠ ⚠

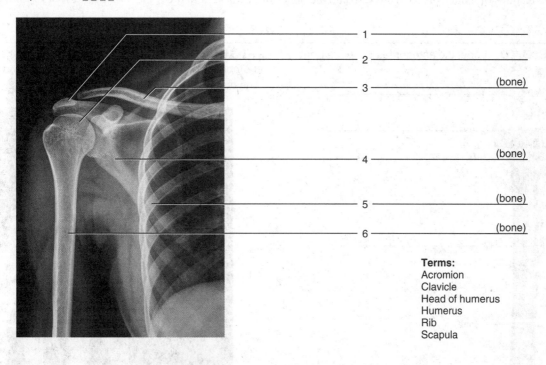

1 _____

2 _____

3 _____ (bone)

4 _____ (bone)

5 _____ (bone)

6 _____ (bone)

Terms:
Acromion
Clavicle
Head of humerus
Humerus
Rib
Scapula

FIGURE 8.9 Identify the bones indicated on this radiograph of the right hand (anterior view), using the terms provided. ⚠

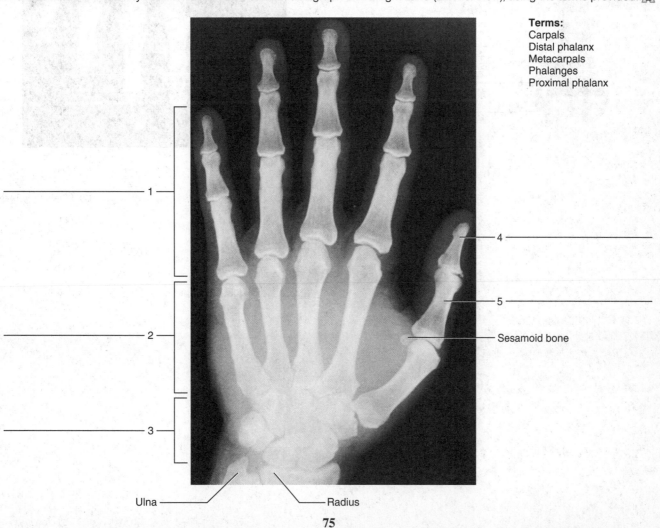

Terms:
Carpals
Distal phalanx
Metacarpals
Phalanges
Proximal phalanx

1 _____

2 _____

3 _____

4 _____

5 _____

Sesamoid bone

Ulna _____   Radius _____

Part E Assessments

Identify the features of a humerus in figure 8.10 and the bones of the hand in figure 8.11.

FIGURE 8.10 Label the anterior features of a right humerus. 🄰

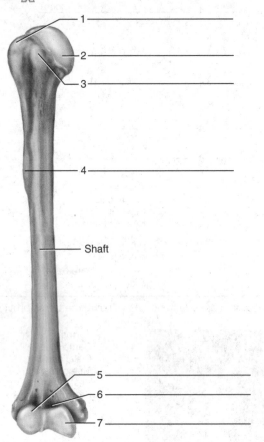

Shaft

FIGURE 8.11 Complete the labeling of the bones numbered on this anterior view of the right hand by placing the correct numbers in the spaces provided. 🄰

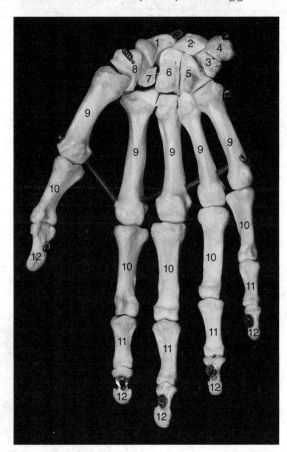

_____	Capitate	___4___	Pisiform
_____	Distal phalanges	_____	Proximal phalanges
_____	Hamate	_____	Scaphoid
_____	Lunate	_____	Trapezium
_____	Metacarpals	_____	Trapezoid
_____	Middle phalanges	___3___	Triquetrum

Pelvic Girdle and Lower Limb

Purpose of the Exercise

To examine the bones of the pelvic girdle and lower limb, and to identify the major features of these bones.

Materials Needed

Human skeleton, articulated
Human skeleton, disarticulated
Male and female pelves

Learning Outcomes

After completing this exercise, you should be able to

1. Locate and identify the bones of the pelvic girdle and their major features.

2. Differentiate a male and female pelvis.

3. Locate and identify the bones of the lower limb and their major features.

Pre-Lab

Carefully read the introductory material and examine the entire lab. Be familiar with the pelvic girdle and the lower limb bones from lecture or the textbook. Answer the pre-lab questions.

Pre-Lab Questions: Select the correct answer for each of the following questions:

1. The two hip bones articulate anteriorly at the
   a. acetabulum.   **b.** pubic arch.
   c. sacroiliac joint.   **d.** pubic symphysis.

2. Anatomically, *leg* refers to
   a. the lower limb.   **b.** hip to knee.
   c. knee to ankle.   **d.** hip to ankle.

3. The _____ is the largest portion of the hip bone.
   a. acetabulum   **b.** ilium
   c. ischium   **d.** pubis

4. The _____ is the lateral bone in the leg.
   a. tibia   **b.** fibula
   c. femur   **d.** patella

5. Which of the following bones is *not* a tarsal bone?
   a. metatarsal   **b.** talus
   c. calcaneus   **d.** cuboid

6. The ilium, ischium, and pubis are separate bones in a young child.
   True _____   False _____

7. Ischial spines, ischial tuberosities, and iliac crests are closer together in a pelvis of a female than in a pelvis of a male.
   True _____   False _____

8. Each digit of a foot has three phalanges.
   True _____   False _____

The pelvic girdle includes two hip bones that articulate with each other anteriorly at the pubic symphysis. Posteriorly, each hip bone articulates with a sacrum at a sacroiliac joint. Together, the pelvic girdle, sacrum, and coccyx comprise the pelvis. The pelvis, in turn, provides support for the trunk of the body and provides attachments for the lower limbs. The pelvis supports and protects the viscera in the pelvic region of the abdominopelvic cavity. The pelvic outlet, with boundaries of the coccyx, inferior border of the pubic symphysis, and between the ischial tuberosities, is clinically important in females. The pelvic outlet must be large enough to successfully accommodate the fetal head during a vaginal delivery. Each acetabulum of a hip bone articulates with the head of the femur of a lower limb. The hip joint structures provide a more stable joint compared to a shoulder joint.

The bones of the lower limb form the framework of the thigh, leg, and foot. (Anatomically, *thigh* represents the region from hip to knee, *leg* is from knee to ankle, and *foot* includes the ankle to the end of the toes.) Each limb includes a femur in the thigh, a patella in the knee, a tibia and fibula in the leg, and seven tarsals, five metatarsals, and fourteen phalanges in the foot. These bones and large muscles are for weight-bearing support and locomotion and thus are considerably larger and possess more stable joints than those of an upper limb.

Procedure A—Pelvic Girdle

The pelvic girdle supports the majority of the weight of the head, neck, and trunk. Each hip bone (coxal bone) originates in three separate ossification areas known as the ilium, ischium, and pubis. These are separate bones of a child but fuse into an individual hip bone. All three parts of the hip bone fuse within the acetabulum, and the ischium and pubis also fuse along the inferior portion of the obturator foramen. The acetabulum is a well-formed deep socket for the head of the femur. The obturator foramen is the largest foramen in the skeleton; it serves as a passageway for blood vessels and nerves between the pelvic cavity and the thigh.

Several of the bone features (bone markings) of the pelvis can be palpated because they are located near the surface of the body. The medial sacral crest is near the middle of the posterior surface of the pelvis, somewhat superior to the coccyx. The ilium is the largest portion of the hip bone, and its iliac crest can be palpated along the anterior and lateral portions. By following the anterior portion of the iliac crest, the anterior superior iliac spine can be felt very close to the surface.

1. Examine figures 9.1 and 9.2.
2. Observe the bones of the pelvic girdle and locate the following:

   hip bone (coxal bone; pelvic bone; innominate bone)
   - ilium
     - iliac crest
     - anterior superior iliac spine
     - anterior inferior iliac spine
     - posterior superior iliac spine
     - posterior inferior iliac spine
     - greater sciatic notch—portion in ischium
     - iliac fossa

FIGURE 9.1 Bones of the pelvis (anterosuperior view).

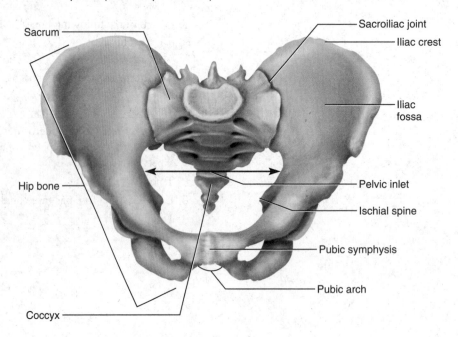

Sacrum

Sacroiliac joint

Iliac crest

Iliac fossa

Hip bone

Pelvic inlet

Ischial spine

Pubic symphysis

Pubic arch

Coccyx

FIGURE 9.2 Right hip bone (a) lateral view (b) medial view. The three colors used enable the viewing of fusion locations of the ilium, ischium, and pubis of the adult skeleton.

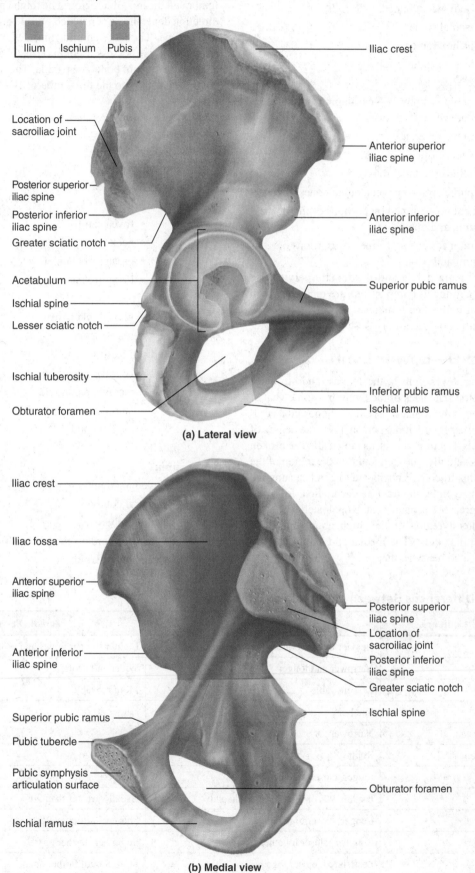

Ilium   Ischium   Pubis

Location of sacroiliac joint

Posterior superior iliac spine

Posterior inferior iliac spine

Greater sciatic notch

Acetabulum

Ischial spine

Lesser sciatic notch

Ischial tuberosity

Obturator foramen

Iliac crest

Anterior superior iliac spine

Anterior inferior iliac spine

Superior pubic ramus

Inferior pubic ramus

Ischial ramus

(a) Lateral view

Iliac crest

Iliac fossa

Anterior superior iliac spine

Anterior inferior iliac spine

Superior pubic ramus

Pubic tubercle

Pubic symphysis articulation surface

Ischial ramus

Posterior superior iliac spine

Location of sacroiliac joint

Posterior inferior iliac spine

Greater sciatic notch

Ischial spine

Obturator foramen

(b) Medial view

- ischium
    - ischial tuberosity—supports weight of body when seated
    - ischial spine
    - ischial ramus
    - lesser sciatic notch
- pubis
    - pubic symphysis—cartilaginous joint between pubic bones
    - pubic tubercle
    - superior pubic ramus
    - inferior pubic ramus
    - pubic arch—between pubic bones of pelvis
- acetabulum—formed by portions of ilium, ischium, and pubis
- obturator foramen—formed by portions of ischium and pubis

3. Observe the male pelvis and the female pelvis. Use table 9.1 as a guide as comparisons are made between the pelves of males and females.
4. Complete Part A of Laboratory Assessment 9.

Procedure B—Lower Limb

The femur of the lower limb is the longest and strongest bone of the skeleton. The neck of the femur has somewhat of a lateral angle and is the weakest part of the bone and a common fracture site, especially if the person has some degree of osteoporosis. This fracture site is usually called a broken hip; however it is actually a broken femur in the region of the hip joint. Often this fracture is attributed to a fall causing the fracture, when many times the fracture occurs first, followed by the fall. The greater trochanter can be palpated along the proximal and lateral region of the thigh about one hand length below the iliac crest. The medial and lateral epicondyles can be palpated near the knee.

Large muscles are positioned on the anterior and posterior surfaces as well as the lateral and medial surfaces of the femur and the hip joint region. The thigh can be pulled in any direction depending upon which muscle group is contracting at that time. The large anterior muscles of the thigh (quadriceps femoris) possess a common patellar tendon with an enclosed sesamoid bone, the patella. The muscle attachment continues distally to the tibial tuberosity.

1. Examine figures 9.3, 9.4, and 9.5.
2. Observe the bones of the lower limb and locate each of the following:

femur
- proximal features
    - head
    - fovea capitis
    - neck
    - greater trochanter
    - lesser trochanter
- shaft
    - gluteal tuberosity
    - linea aspera
- distal features
    - lateral epicondyle
    - medial epicondyle
    - lateral condyle
    - medial condyle

patella

tibia
- medial condyle
- lateral condyle
- tibial tuberosity
- anterior border (crest; margin)
- medial malleolus

TABLE 9.1 Differences Between Male and Female Pelves

Structure of Comparison	Male Pelvis	Female Pelvis
General structure	Heavier thicker bones and processes	Lighter thinner bones and processes
Sacrum	Narrower and longer	Wider and shorter
Coccyx	Less movable	More movable
Pelvic outlet	Smaller	Larger
Greater sciatic notch	Narrower	Wider
Obturator foramen	Round	Triangular to oval
Acetabula	Larger; closer together	Smaller; farther apart
Pubic arch	Usually 90° or less; more V-shaped	Usually greater than 90°
Ischial spines	Longer; closer together	Shorter; farther apart
Ischial tuberosities	Rougher; closer together	Smoother; farther apart
Iliac crests	Less flared; closer together	More flared; farther apart

FIGURE 9.3 The features of (a) the anterior surface and (b) the posterior surface of the right femur. The anterior and posterior views of the patella are included in the figure.

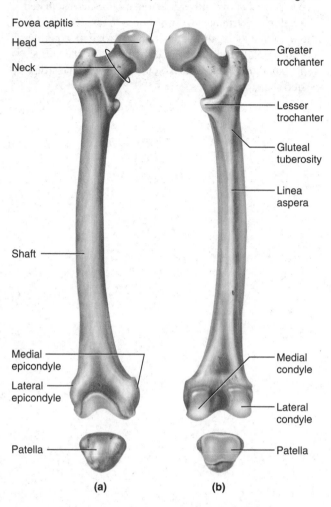

FIGURE 9.4 Features of the right tibia and fibula (anterior view).

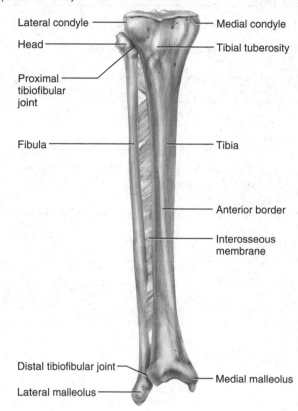

FIGURE 9.5 Superior surface view of right foot. The proximal group of tarsals are colored yellow and the distal group of tarsals are colored green.

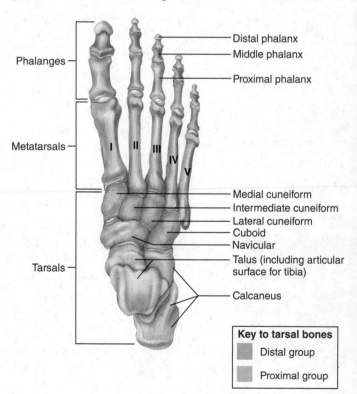

fibula
- head
- lateral malleolus

tarsal bones
- talus
- calcaneus
- navicular
- cuboid
- lateral cuneiform
- intermediate (middle) cuneiform
- medial cuneiform

metatarsal bones

phalanges
- proximal phalanx
- middle phalanx—absent in first digit
- distal phalanx

3. Complete Parts B, C, and D of the laboratory assessment.

Critical Thinking Activity

Compare the size and depth of an acetabulum of a hip bone to the glenoid cavity of the scapula._____

_____

_____

_____

_____

How do these socket differences between the hip joint and the shoulder joint relate to strength and mobility?

_____

_____

_____

_____

_____

Critical Thinking Activity

As a review of the entire skeleton, use the disarticulated skeleton and arrange all of the bones in relative position to re-create their normal positions. Work with a partner or in a small group and place the bones on the surface of a laboratory table. Check your results with the articulated skeletons.

Laboratory Assessment

9

Name _____

Date _____

Section _____

The Ⓐ corresponds to the indicated outcome(s) found at the beginning of the laboratory exercise.

Pelvic Girdle and Lower Limb

Part A Assessments

Complete the following statements:

1. The pelvic girdle consists of two _____. Ⓐ

2. The head of the femur articulates with the _____ of the hip bone. Ⓐ

3. The _____ is the largest portion of the hip bone. Ⓐ

4. The distance between the _____ represents the shortest diameter of the pelvic outlet. Ⓐ

5. The pubic bones come together anteriorly to form a cartilaginous joint called the _____. Ⓐ

6. The _____ is the superior margin of the ilium that causes the prominence of the hip. Ⓐ

7. When a person sits, the _____ of the ischium supports the weight of the body. Ⓐ

8. The angle formed by the pubic bones below the pubic symphysis is called the _____. Ⓐ

9. The _____ is the largest foramen in the skeleton. Ⓐ

10. The ilium joins the sacrum at the _____ joint. Ⓐ

Critical Thinking Assessment

Examine the male and female pelves. Look for major differences between them. Note especially the flare of the iliac bones, the angle of the pubic arch, the distance between the ischial spines and ischial tuberosities, and the curve and width of the sacrum. In what ways are the differences you observed related to the function of the female pelvis as a birth canal? Ⓐ

_____

_____

_____

Part B Assessments

Match the bones in column A with the features in column B. Place the letter of your choice in the space provided. Ⓐ

Column A	Column B	
a. Femur	_____ 1. Middle phalanx	_____ 7. Tibial tuberosity
b. Fibula	_____ 2. Lesser trochanter	_____ 8. Talus
c. Metatarsals	_____ 3. Medial malleolus	_____ 9. Linea aspera
d. Patella	_____ 4. Fovea capitis	_____ 10. Lateral malleolus
e. Phalanges	_____ 5. Calcaneus	_____ 11. Sesamoid bone
f. Tarsals	_____ 6. Lateral cuneiform	_____ 12. Five bones that form the instep
g. Tibia		

Part C Assessments

Identify the bones and features indicated in the radiographs of figures 9.6, 9.7, and 9.8.

FIGURE 9.6 Identify the bones and features indicated on this radiograph of the anterior view of the pelvic region, using the terms provided. /A\ /3\

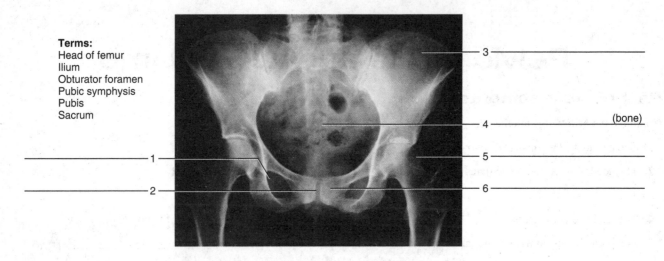

Terms:
Head of femur
Ilium
Obturator foramen
Pubic symphysis
Pubis
Sacrum

3 _____

4 _____ (bone)

5 _____

6 _____

1 _____

2 _____

FIGURE 9.7 Identify the bones and features indicated in this radiograph of the right knee (anterior view), using the terms provided. /3\

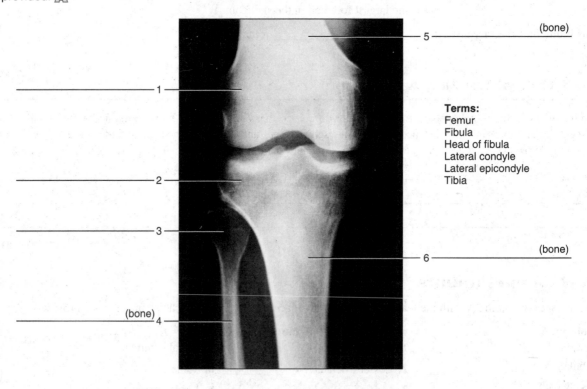

5 _____ (bone)

Terms:
Femur
Fibula
Head of fibula
Lateral condyle
Lateral epicondyle
Tibia

6 _____ (bone)

1 _____

2 _____

3 _____

(bone) 4 _____

FIGURE 9.8 Identify the bones indicated in this radiograph of the right foot (medial side), using the terms provided. ⚠

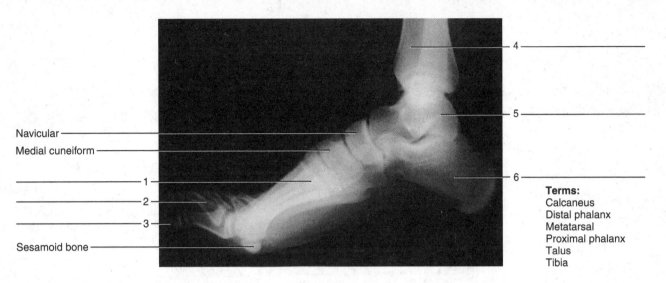

Navicular

Medial cuneiform

1

2

3

Sesamoid bone

4

5

6

Terms:
Calcaneus
Distal phalanx
Metatarsal
Proximal phalanx
Talus
Tibia

Part D Assessments

Identify the bones of the foot in figure 9.9 and the features of a femur in figure 9.10.

FIGURE 9.9 Identify the bones indicated on this superior view of the right foot, using the terms provided. ⚠

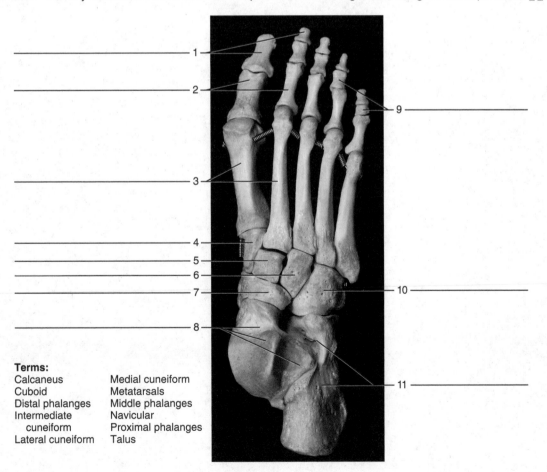

1

2

3

4

5

6

7

8

9

10

11

Terms:
Calcaneus   Medial cuneiform
Cuboid   Metatarsals
Distal phalanges   Middle phalanges
Intermediate   Navicular
  cuneiform   Proximal phalanges
Lateral cuneiform Talus

FIGURE 9.10 Label the anterior features of a right femur. 3

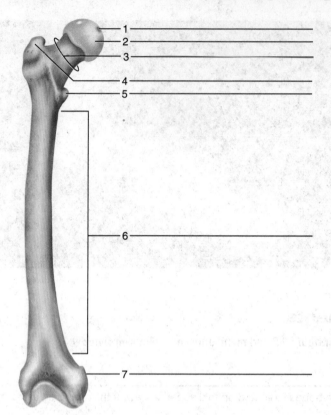

1 _____

2 _____

3 _____

4 _____

5 _____

6 _____

7 _____

Skeletal Muscle Structure and Function

Purpose of the Exercise

To study the structure and function of skeletal muscles as cells and as organs.

Materials Needed

Compound light microscope
Prepared microscope slide of skeletal muscle tissue
  (longitudinal section and cross section)
Human torso model with musculature
Model of skeletal muscle fiber

For Demonstration Activity:
Fresh round beefsteak

Learning Outcomes

After completing this exercise, you should be able to

1. Locate the structures of a skeletal muscle fiber (cell).
2. Describe how connective tissue is associated with muscle tissue within a skeletal muscle.
3. Distinguish between the origin and insertion of a muscle.
4. Describe and demonstrate the general actions of prime movers (agonists), synergists, fixators, and antagonists.
5. Diagram and label a muscle twitch.
6. Demonstrate *threshold* (the minimum voltage required to observe the appearance of muscle contraction).
7. Illustrate *recruitment* (the stimulation of additional motor units in the muscle) by observing the change in amplitude of the contraction.
8. Determine maximum contraction (the stimulation of all motor units in the muscle) by observing no further increase in the amplitude of the contraction.
9. Measure amplitude of muscle contraction when the muscle is stimulated with varying degrees of voltage.
10. Integrate the concepts of recruitment and maximum stimulation as applied to the muscles of the body.

Pre-Lab

Carefully read the introductory material and examine the entire lab. Be familiar with skeletal muscle tissue and muscle structure and function from lecture or the textbook. Answer the pre-lab questions.

Pre-Lab Questions: Select the correct answer for each of the following questions:

1. The outermost layer of connective tissue of a muscle is the
   a. fascicle.   **b.** endomysium.
   c. epimysium.   **d.** perimysium.
2. The thick myofibril filament of a sarcomere is composed of a protein
   a. myosin.   **b.** actin.
   c. titin.   **d.** sarcolemma.
3. The muscle primarily responsible for a movement is the
   a. synergist.   **b.** prime mover (agonist).
   c. antagonist.   **d.** origin.
4. The neuron and the collection of muscle fibers it innervates is called the
   a. neural impulse. **b.** muscle fiber.
   c. motor neuron.   **d.** motor unit.
5. The functional contractile unit of a muscle fiber (cell) is a
   a. sarcoplasm.   **b.** sarcolemma.
   c. sarcomere.   **d.** sarcoplasmic reticulum.
6. The role of a particular muscle is always the same.
   True _____   False _____
7. A synergistic muscle contraction assists the prime mover.
   True _____   False _____
8. The plasma membrane of a muscle fiber (cell) is called the sarcolemma.
   True _____   False _____

A skeletal muscle represents an organ of the muscular system and is composed of several types of tissues. These tissues include skeletal muscle tissue, nervous tissue, and various connective tissues.

Each skeletal muscle is encased and permeated with connective tissue sheaths. The connective tissues surround and extend into the structure of a muscle and separate it into compartments. The entire muscle is encased by the *epimysium*. The *perimysium* covers bundles of cells (*fascicles*), within the muscle. The deepest connective tissue surrounds each individul muscle fiber (cell) as a thin *endomysium*. The connective tissues provide support and reinforcement during muscular contractions and allow portions of a muscle to contract somewhat independently. The connective tissue often extends beyond the end of a muscle, providing an attachment to other muscles or to bones. Some collagen fibers of the connective tissue are continuous with the tendon and the periosteum, making for a strong structural continuity.

Muscles are named according to their location, size, shape, action, attachments, number of origins, or the direction of the fibers. Examples of how muscles are named include: gluteus maximus (location and size); adductor longus (action and size); sternocleidomastoid (attachments); serratus anterior (shape and location); biceps (two origins); and orbicularis oculi (direction of fibers and location).

Skeletal muscles, such as the biceps brachii, are composed of many muscle fibers (cells). These muscle fibers are innervated by motor neurons of the nervous system. Anatomically, each motor neuron is arranged to extend to a specific number of muscle fibers. (Note that each muscle fiber is innervated by only one motor neuron.) The motor neuron and the collection of muscle fibers it innervates is called a *motor unit*. When the motor neuron relays a neural impulse to the muscle, all the muscle fibers that it innervates (i.e., the motor unit) will be stimulated to contract. If only a few motor units are stimulated, the muscle as a whole exerts little force. If a larger number of motor units are stimulated, the muscle will exert a greater force. Thus, to lift a heavy object like a suitcase requires the activation of more motor units in the biceps brachii than the lifting of a lighter object like a book.

Procedure A—Skeletal Muscle Structure

1. Examine the microscopic structure of a longitudinal section of skeletal muscle by observing a prepared microscope slide of this tissue. Use figure 10.1 of skeletal muscle tissue to locate the following features:

    skeletal muscle fiber (cell)

    nuclei

    striations (alternating light and dark)

2. Skeletal muscle cells are stimulated by nerve impulses over cellular processes (axons) of motor neurons. The *neuromuscular junctions* are the sites where the axons terminate at a muscle fiber (fig. 10.1).

3. Study figures 10.2 and 10.3. Note the arrangement of muscle fibers in relation to the connective tissues of the

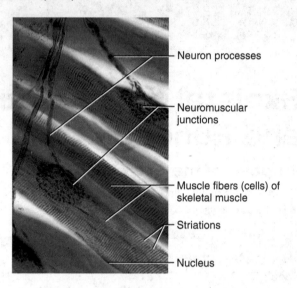

FIGURE 10.1 Micrograph of a longitudinal section of skeletal muscle fibers with associated neuromuscular junctions (400×).

- Neuron processes
- Neuromuscular junctions
- Muscle fibers (cells) of skeletal muscle
- Striations
- Nucleus

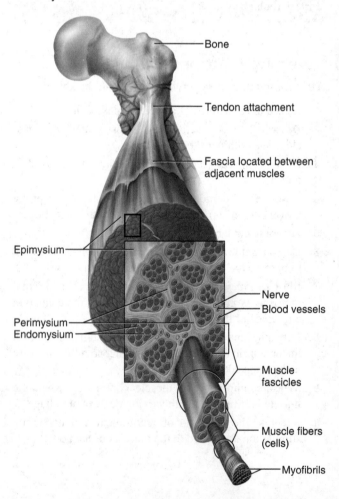

FIGURE 10.2 Skeletal muscle structure from the gross anatomy to the microscopic arrangement. Note the distribution pattern of the epimysium, perimysium, and endomysium.

- Bone
- Tendon attachment
- Fascia located between adjacent muscles
- Epimysium
- Nerve
- Blood vessels
- Perimysium
- Endomysium
- Muscle fascicles
- Muscle fibers (cells)
- Myofibrils

FIGURE 10.3 Cross section of a fascicle and associated connective tissues (scanning electron micrograph, 320×).

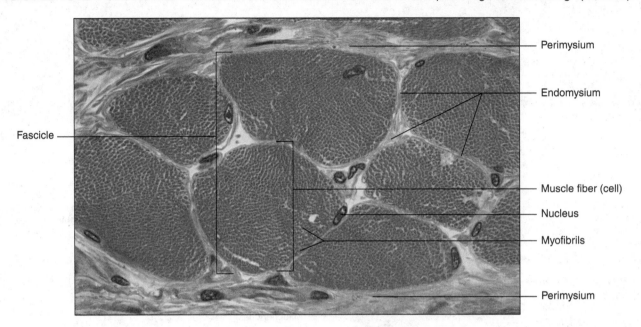

Fascicle

Perimysium

Endomysium

Muscle fiber (cell)

Nucleus

Myofibrils

Perimysium

epimysium, perimysium, and endomysium of an individual muscle. Examine the microscopic structure of a cross section of skeletal muscle tissue using figure 10.3 for identification of the structures.

4. Examine the human torso model and locate examples of tendons and aponeuroses. An *origin tendon* is attached to a fixed location, while the *insertion tendon* is attached to a more movable location. Sheets of connective tissue, called aponeuroses, also serve for some muscle attachments. An example of a large aponeurosis visible on the torso or a muscle chart is in the abdominal area. It appears as a broad white sheet of connective tissue for some of the abdominal muscle attachments to each other. Locate examples of cordlike tendons in your body. The large calcaneal tendon is easy to palpate in a posterior ankle just above the heel.

 Demonstration Activity

Examine the fresh round beefsteak. It represents a cross section through the beef thigh muscles. Note the white lines of connective tissue that separate the individual skeletal muscles. Also note how the connective tissue extends into the structure of a muscle and separates it into small compartments of muscle tissue. Locate the epimysium and the perimysium of an individual muscle.

 Safety

▶ Wear disposable gloves when handling the fresh beefsteak.
▶ Wash your hands before leaving the laboratory.

FIGURE 10.4 Structures of a segment of a muscle fiber (cell).

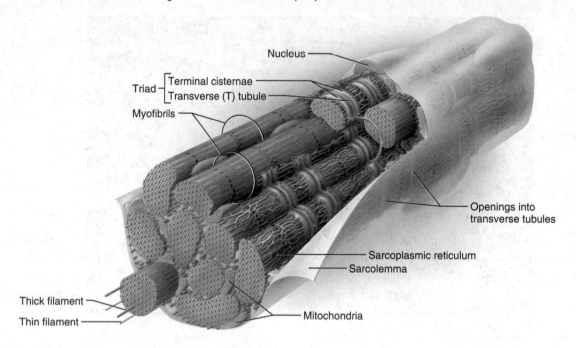

Nucleus
Triad — Terminal cisternae
Transverse (T) tubule
Myofibrils
Openings into transverse tubules
Sarcoplasmic reticulum
Sarcolemma
Thick filament
Thin filament
Mitochondria

5. Using figures 10.4 and 10.5 as a guide, examine the model of the skeletal muscle fiber. Locate the following:

sarcolemma—plasma membrane of a muscle fiber

sarcoplasm—cytoplasm of a muscle fiber

myofibril—bundle of protein filaments

- thick filament—composed of contractile protein myosin
- thin filament—composed mostly of contractile protein actin
- elastic filament—composed of springy protein titin

sarcoplasmic reticulum (SR)—smooth ER of muscle fiber

- terminal cisternae—extended ends of SR next to transverse tubules

transverse (T) tubules—inward extensions of sarcolemma through the muscle fiber

sarcomere—functional contractile unit within muscle fiber (fig. 10.5)

- A (anisotropic) band—dark band of thick filaments and some overlap of thin filaments
- I (isotropic) band—light band of thin filaments
- H zone (band)—light band in middle of A band
- M line—dark line in middle of H zone
- Z disc (line)—thin and elastic filaments anchored at ends of sarcomere

FIGURE 10.5 Sarcomere contractile unit of a muscle fiber: (a) micrograph (16,000×); (b) illustration of the repeating pattern of striations (bands).

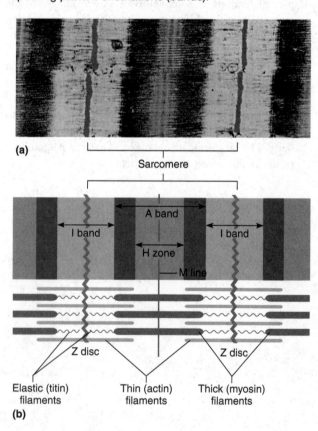

(a)

Sarcomere

A band
I band
H zone
M line
I band

Z disc
Z disc

Elastic (titin) filaments
Thin (actin) filaments
Thick (myosin) filaments

(b)

FIGURE 10.6 Antagonistic muscle pairs are located on opposite sides of the same body region as shown in an arm. The muscle acting as the prime mover depends upon the movement that occurs.

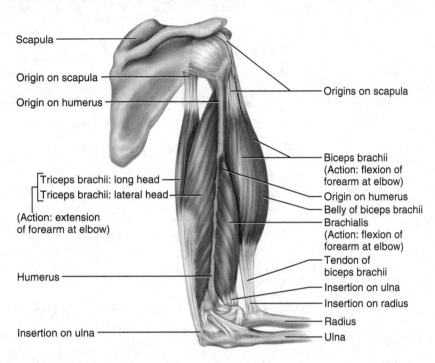

TABLE 10.1 Various Roles of Muscles

Functional Category	Description
Prime mover (agonist)	Muscle primarily responsible for the action (movement)
Antagonist	Muscle responsible for action in the opposite direction of a prime mover or for resistance to a prime mover
Synergist	Muscle contraction assists a prime mover
Fixator	Special type of synergist muscle; muscle contraction will stabilize a joint so another contracting muscle exerts a force on something else

6. Complete Part A and B of the Laboratory Assessment 10.
7. Study figure 10.6 and table 10.1.
8. Locate the biceps brachii, brachialis, and triceps brachii and their origins and insertions in the human torso model and in your body.
9. Make various movements with your upper limb at the shoulder and elbow. For each movement, determine the location of the muscles functioning as prime movers (agonists) and as antagonists. When a prime mover contracts (shortens) and a joint moves, its antagonist relaxes (lengthens). Antagonistic muscle pairs pull from opposite sides of the same body region. Synergistic muscles often supplement the contraction force of a prime mover, or by acting as fixators, they might also stabilize nearby joints. **The role of a muscle as a prime mover, antagonist, or synergist depends upon the movement under consideration, as their roles change.**
10. Complete Part C of the laboratory assessment.

Name _____

Date _____

Section _____

The ⚠ corresponds to the indicated outcome(s) found at the beginning of the laboratory exercise.

Skeletal Muscle Structure and Function

Part A Assessments

Match the terms in column A with the definitions in column B. Place the letter of your choice in the space provided. ⚠ ⚠

Column A	Column B
a. Endomysium	_____ **1.** Membranous channel extending inward from muscle fiber membrane
b. Epimysium	_____ **2.** Cytoplasm of a muscle fiber
c. Fascia	_____ **3.** Connective tissue located between adjacent muscles
d. Fascicle	_____ **4.** Layer of connective tissue that separates a muscle into small bundles called fascicles
e. Myosin	
f. Perimysium	_____ **5.** Plasma membrane of a muscle fiber
g. Sarcolemma	_____ **6.** Layer of connective tissue that surrounds a skeletal muscle
h. Sarcomere	_____ **7.** Unit of alternating light and dark striations between Z discs (lines)
i. Sarcoplasm	
j. Sarcoplasmic reticulum	_____ **8.** Layer of connective tissue that surrounds an individual muscle fiber
k. Tendon	_____ **9.** Cellular organelle in muscle fiber corresponding to the endoplasmic reticulum
l. Transverse (T) tubule	
	_____ **10.** Cordlike part that attaches a muscle to a bone
	_____ **11.** Protein found within thick filament
	_____ **12.** A small bundle of muscle fibers within a muscle

Part B Assessments

Provide the labels for the electron micrograph in figure 10.7.

FIGURE 10.7 Label this transmission electron micrograph (16,000×) of a relaxed sarcomere by placing the correct numbers in the spaces provided. ⚠

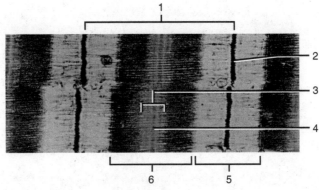

___ A band (dark)
___ H zone
___ I band (light)
___ M line
___ Sarcomere
___ Z disc

Part C   Assessments

Complete the following statements:

1. The _____ of a muscle is usually attached to a fixed location. ⒊

2. The _____ of a muscle is usually attached to a movable location. ⒊

3. A muscle responsible for most of a movement is called a(n) _____. ⒋

4. Assisting muscles are called _____. ⒋

5. Antagonists are muscles that resist the actions of _____ and cause movement in the opposite direction. ⒋

6. When the forearm is extended at the elbow joint, the _____ muscle acts as the prime mover. ⒋

7. When the biceps brachii acts as the prime mover, the _____ muscle assists as a synergist. ⒋

Critical Thinking Assessment

You are getting ready to leave the lab. You lift your pencil to put it in your book bag. Then you lift your lab book and place it in your book bag. Which scenario resulted from the stimulation of more motor units? ⒑

_____

Which would require more neural signals being sent by the nervous system to the muscles of your arm? ⒑

_____

What concept covered in this lab (threshold, recruitment, or maximum contraction) explains why there are certain objects that are too heavy for you to lift? ⒑

_____

_____

Muscles of the Head and Neck

Purpose of the Exercise

To review the locations, actions, origins, and insertions of the muscles of the head and neck.

Materials Needed

Human torso model with musculature
Human skull
Human skeleton, articulated

For Learning Extension Activity:
Long rubber bands

Learning Outcomes

After completing this exercise, you should be able to

1. Locate and identify the muscles of facial expression, the muscles of mastication, the muscles that move the head and neck, and the muscles that move the hyoid bone and larynx.

2. Describe and demonstrate the action of each of these muscles.

3. Locate the origin and insertion of each of these muscles in a human skeleton and the musculature of the human torso model.

Pre-Lab

Carefully read the introductory material and examine the entire lab. Be familiar with the locations, actions, origins, and insertions of the muscles of the head and neck from lecture or the textbook. Answer the pre-lab questions.

Pre-Lab Questions: Select the correct answer for each of the following questions:

1. Muscles of mastication are all inserted on the
   a. mandible. b. maxilla.
   c. tongue. d. teeth.

2. Muscles of mastication include the following *except* the
   a. lateral pterygoid. b. masseter.
   c. temporalis. d. platysma.

3. Facial expressions are so variable partly because many facial muscles are inserted
   a. on the mandible. b. in the skin.
   c. on the maxilla. d. on facial bones.

4. Which of the following muscles does *not* move the hyoid bone?
   a. sternocleidomastoid b. mylohyoid
   c. sternohyoid d. thyrohyoid

5. Which of the following is *not* a facial expression muscle?
   a. zygomaticus major b. buccinator
   c. orbicularis oris d. scalenes

6. Flexion of the head and neck is an action of the _____ muscle.
   a. semispinalis capitis b. splenius capitis
   c. sternocleidomastoid d. trapezius

7. Which of the following muscles is the most visible from an anterior view of a person in anatomical position?
   a. splenius capitis b. nasalis
   c. temporalis d. medial pterygoid

The skeletal muscles of the head include the muscles of facial expression. Human facial expressions are more variable than those of other mammals because many of the muscles have insertions in the skin, rather than on bones, allowing great versatility of movements and emotions. The extrinsic muscles of eye movements will be covered in a later laboratory exercise.

The muscles of mastication (chewing) are all inserted on the mandible. Muscles for chewing allow the movements of depression and elevation of the mandible to open and close the jaw. However, jaw movements also involve protraction, retraction, lateral excursion, and medial excursion needed to bite off pieces of food and grind the food.

The muscles that move the head are located in the neck. The actions of flexion, extension, and hyperextension of the head and the cervical vertebral column occur when right and left paired muscles work simultaneously. Additional actions, such as lateral flexion and rotation, result from muscles on one side only contracting or alternating muscles contracting.

A group of neck muscles that have attachments on the hyoid bone and larynx assist in swallowing and speech. Some of the neck muscles that move the head and neck have their attachments in the thorax.

Procedure—Muscles of the Head and Neck

The muscle lists and tables in this laboratory exercise reflect muscle groupings using a combination of related locations and functions. Study and master one group at a time as a separate task. Frequently refer to the illustrations, tables, skeletons, and models. Consider working with a partner or a small group to study and review.

1. Study figure 11.1 and table 11.1.
2. Locate the following muscles in the human torso model and in your body whenever possible:

   muscles of facial expression
   - epicranius (occipitofrontalis)
     - frontal belly (frontalis)
     - occipital belly (occipitalis)
   - nasalis
   - orbicularis oculi
   - orbicularis oris
   - risorius
   - zygomaticus major
   - zygomaticus minor
   - buccinator
   - platysma

3. Demonstrate the actions of these muscles in your body.
4. Locate the origins and insertions of these muscles in the human skull and skeleton.
5. Study figures 11.1 and 11.2 and table 11.2.
6. Locate the following muscles in the human torso model and in your body whenever possible:

   muscles of mastication
   - masseter
   - temporalis
   - lateral pterygoid
   - medial pterygoid

FIGURE 11.1 Muscles of facial expression and mastication (a) anterior view and (b) lateral view.

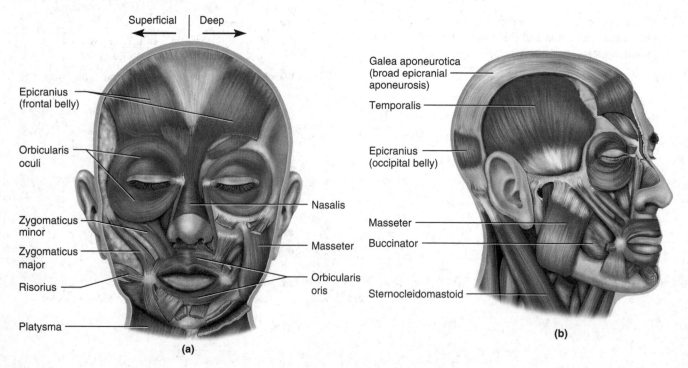

(a)

(b)

TABLE 11.1 Muscles of Facial Expression

Muscle	Origin	Insertion	Action
Epicranius	Occipital bone; galea aponeurotica	Skin of eyebrows; galea aponeurotica	Raises eyebrows; retracts scalp
Nasalis	Maxilla lateral to nose	Bridge of nose	Widens nostrils
Orbicularis oculi	Maxillary and frontal bones	Skin around eye	Closes eyes as in blinking
Orbicularis oris	Muscles near the mouth	Skin of lips	Closes lips; protrudes lips as for kissing
Risorius	Fascia near ear	Corner of mouth	Draws corner of mouth laterally
Zygomaticus major	Zygomatic bone	Corner of mouth	Raises corner of mouth as when smiling and laughing
Zygomaticus minor	Zygomatic bone	Corner of mouth	Raises corner of mouth as when smiling and laughing
Buccinator	Lateral surfaces of maxilla and mandible	Orbicularis oris	Compresses cheeks inward as when blowing air
Platysma	Fascia in upper chest	Lower border of mandible and skin at corner of mouth	Draws angle of mouth downward as when pouting or expressing horror

FIGURE 11.2 Deep muscles of mastication.

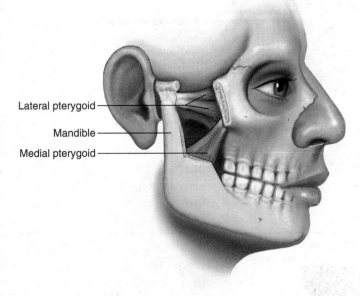

Lateral pterygoid

Mandible

Medial pterygoid

TABLE 11.2 Muscles of Mastication

Muscle	Origin	Insertion	Action
Masseter	Lower border of zygomatic arch	Lateral surface of mandible	Elevates mandible
Temporalis	Temporal bone	Coronoid process and anterior ramus of mandible	Elevates mandible
Lateral pterygoid	Sphenoid bone	Anterior surface of mandibular condyle	Depresses and protracts mandible and moves it from side to side as when grinding food
Medial pterygoid	Sphenoid, palatine, and maxillary bones	Medial surface of mandible	Elevates mandible and moves it from side to side as when grinding food

7. Demonstrate the actions of these muscles in your body.
8. Locate the origins and insertions of these muscles in the human skull and skeleton.
9. Study figures 11.3 and 11.4 and table 11.3.
10. Locate the following muscles in the human torso model and in your body whenever possible:

   muscles that move head and neck
   - sternocleidomastoid
   - trapezius (superior part)
   - scalenes (anterior, middle, and posterior)
   - splenius capitis
   - semispinalis capitis

11. Demonstrate the actions of these muscles in your body.
12. Locate the origins and insertions of these muscles in the human skull and skeleton.
13. Complete Part A of Laboratory Assessment 11.

14. Study figure 11.5 and table 11.4.
15. Locate the following muscles in the human torso model:

   muscles that move hyoid bone and larynx
   - suprahyoid muscles
     - digastric (2 parts)
     - stylohyoid
     - mylohyoid
   - infrahyoid muscles
     - sternohyoid
     - omohyoid (2 parts)
     - sternothyroid
     - thyrohyoid

16. Demonstrate the actions of these muscles in your body.
17. Locate the origins and insertions of these muscles in the human skull and skeleton.
18. Complete Parts B, C, and D of the laboratory assessment.

Learning Extension Activity

A long rubber band can be used to simulate muscle locations, origins, insertions, and actions on the human torso model, the skeleton, or a laboratory partner. Hold one end of the rubber band firmly on the origin location of a muscle, then slightly stretch the rubber band and hold the other end on the insertion site. Allow the insertion end to slowly move toward the origin end to simulate the contraction and action of the muscle.

FIGURE 11.3 Posterior muscles that move the head and neck.

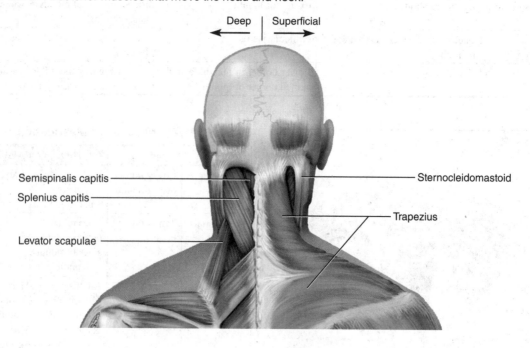

Deep | Superficial

Semispinalis capitis

Splenius capitis

Levator scapulae

Sternocleidomastoid

Trapezius

FIGURE 11.4 Muscles that move the head and neck, lateral view.

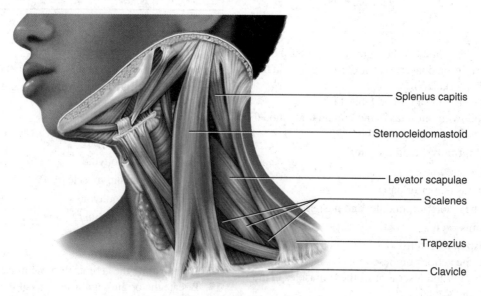

Splenius capitis

Sternocleidomastoid

Levator scapulae

Scalenes

Trapezius

Clavicle

TABLE 11.3   Muscles That Move Head and Neck

Muscle	Origin	Insertion	Action
Sternocleidomastoid	Manubrium of sternum and medial clavicle	Mastoid process of temporal bone	Flexion of head and neck; rotation of head to left or right
Trapezius (superior part)	Occipital bone and spinous processes of C7 and several thoracic vertebrae	Clavicle and the spine and acromion of scapula	Extends head (as a synergist); (primary actions on scapula are covered in another lab)
Scalenes (anterior, middle, and posterior)	Transverse processes of all cervical vertebrae	Ribs 1–2	Elevates ribs 1–2; flexion and rotation of neck
Splenius capitis	Spinous processes of C7–T6	Mastoid process and occipital bone	Extends head; rotates head
Semispinalis capitis	Processes of inferior cervical and superior thoracic vertebrae	Occipital bone	Extends head; rotates head

FIGURE 11.5   Muscles that move the hyoid bone and larynx assist in swallowing and speech and are grouped into suprahyoid and infrahyoid muscles.

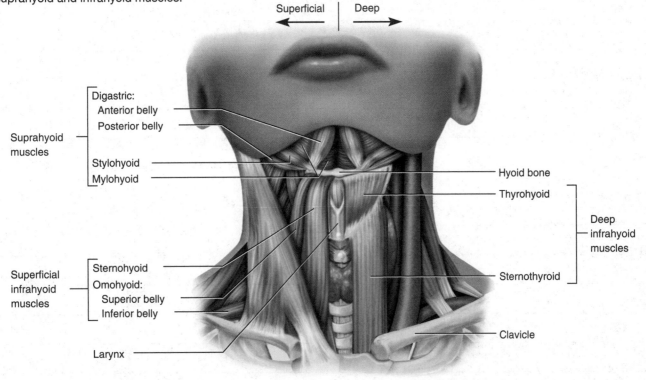

Superficial   Deep

Suprahyoid muscles
- Digastric:
  - Anterior belly
  - Posterior belly
- Stylohyoid
- Mylohyoid

Superficial infrahyoid muscles
- Sternohyoid
- Omohyoid:
  - Superior belly
  - Inferior belly

Larynx

Hyoid bone
Thyrohyoid
Deep infrahyoid muscles
Sternothyroid
Clavicle

TABLE 11.4   Muscles That Move Hyoid Bone and Larynx

Muscle	Origin	Insertion	Action
Digastric (2 parts)	Inferior mandible (anterior belly) and mastoid process (posterior belly)	Hyoid bone	Opens mouth; depresses mandible; elevates hyoid bone
Stylohyoid	Styloid process of temporal bone	Hyoid bone	Retracts and elevates hyoid bone
Mylohyoid	Mandible	Hyoid bone	Elevates hyoid bone during swallowing
Sternohyoid	Manubrium and medial clavicle	Hyoid bone	Depresses hyoid bone
Omohyoid (2 parts)	Superior border of scapula	Hyoid bone	Depresses hyoid bone
Sternothyroid	Manubrium	Thyroid cartilage of larynx	Depresses larynx
Thyrohyoid	Thyroid cartilage of larynx	Hyoid bone	Depresses hyoid bone; elevates larynx

Name _____

Date _____

Section _____

The A corresponds to the indicated outcome(s) found at the beginning of the laboratory exercise.

Muscles of the Head and Neck

Part A Assessments

Identify the muscles indicated in the head and neck in figures 11.6 and 11.7.

FIGURE 11.6   Label the anterior muscles of the head. A

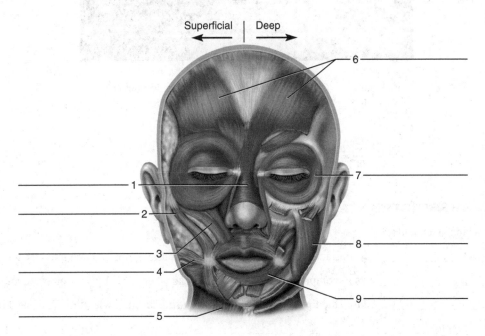

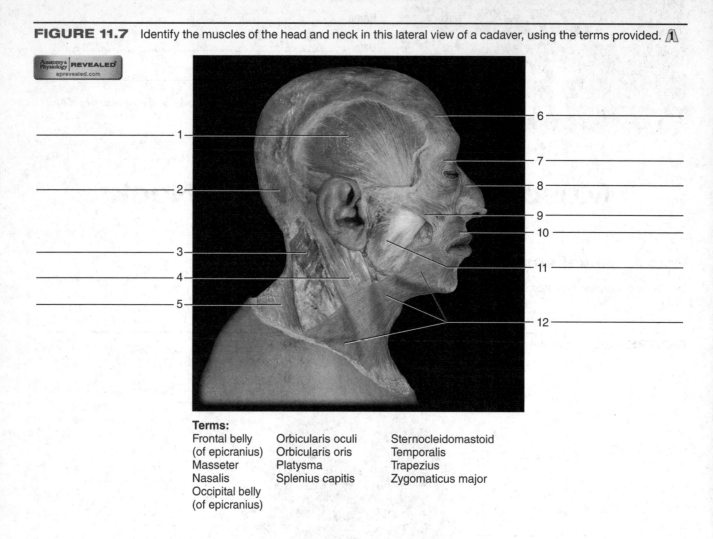

1 _____

2 _____

3 _____

4 _____

5 _____

6 _____

7 _____

8 _____

9 _____

10 _____

11 _____

12 _____

Terms:

Frontal belly	Orbicularis oculi	Sternocleidomastoid
(of epicranius)	Orbicularis oris	Temporalis
Masseter	Platysma	Trapezius
Nasalis	Splenius capitis	Zygomaticus major
Occipital belly		
(of epicranius)		

Part B Assessments

Complete the following statements:

1. When the _____ contracts, the corner of the mouth is drawn upward and laterally when laughing.

2. The _____ acts to compress the wall of the cheeks when air is blown out of the mouth.

3. The _____ causes the lips to close and pucker during kissing, whistling, and speaking.

4. The temporalis acts to _____.

5. The _____ pterygoid can close the jaw and pull it sideways.

6. The _____ pterygoid can protrude the jaw, pull the jaw sideways, and open the mouth.

7. The _____ can close the eye, as in blinking.

8. The _____ can pull the head toward the chest.

9. The muscle used for pouting and to express horror is the _____.

10. The muscle used to widen the nostrils is the _____.

11. The muscle used to elevate the hyoid bone during swallowing is the _____.

12. The _____ raises the eyebrows and moves the scalp.

Muscles of the Hip and Lower Limb

Purpose of the Exercise

To review the actions, origins, and insertions of the muscles that move the thigh, leg, and foot.

Materials Needed

Human torso model with musculature
Human skeleton, articulated
Muscular models of the lower limb
For Learning Extension Activity:
Long rubber bands

Learning Outcomes

After completing this exercise, you should be able to

1. Locate and identify the muscles that move the thigh, leg, and foot.
2. Describe and demonstrate the actions of each of these muscles.
3. Locate the origin and insertion of each of these muscles in a human skeleton and on muscular models.

Pre-Lab

Carefully read the introductory material and examine the entire lab. Be familiar with the locations, actions, origins, and insertions of the muscles of the hip and lower limb from lecture or the textbook. Answer the pre-lab questions.

Pre-Lab Questions: Select the correct answer for each of the following questions:

1. Muscles that move the thigh at the hip joint have origins on the
   a. pelvis.   b. femur.
   c. tibia.   d. patella.
2. Anterior thigh muscles serve as prime movers of
   a. flexion of the leg at the knee joint.
   b. abduction of the thigh at the hip joint.
   c. adduction of the thigh at the hip joint.
   d. extension of the leg at the knee joint.
3. The muscles of the lower limb _____ than those of the upper limb.
   a. are more numerous   b. are larger
   c. conduct actions with   d. are less powerful
      more precision
4. Which of the following muscles is *not* part of the quadriceps group?
   a. rectus femoris   b. vastus lateralis
   c. semitendinosus   d. vastus medialis
5. Which of the following muscles is *not* part of the hamstring group?
   a. rectus femoris   b. biceps femoris
   c. semitendinosus   d. semimembranosus
6. Muscles located in the lateral leg include actions of
   a. extension of the   b. eversion of the foot.
      knee joint.
   c. flexion of the toes. d. extension of the toes.
7. Contractions of the sartorius muscle involve movements across two different joints.
   True _____   False _____
8. The insertions of the four quadriceps femoris muscles are on the femur.
   True _____   False _____

The muscles that move the thigh at the hip joint have their origins on the pelvis and insertions usually on the femur. Those muscles attached on the anterior pelvis act to flex the thigh at the hip joint; those attached on the posterior pelvis act to extend the thigh at the hip joint; those attached on the lateral side of the pelvis act as abductors and rotators of the thigh at the hip joint; and those attached on the medial pelvis act as adductors of the thigh at the hip joint.

The muscles that move the leg at the knee joint have their origins on the femur or the hip bone. The anterior thigh muscles have their insertions on the proximal tibia and serve as prime movers of extension of the leg at the knee joint. The posterior thigh muscles have their insertions on the tibia or fibula and act as prime movers of flexion of the leg at the knee joint.

The muscles that move the foot have their origins on the tibia, fibula, or femur. The anterior leg muscles have actions of dorsiflexion and extension of the toes; the posterior leg muscles have actions of plantar flexion and flexion of the toes; the lateral leg muscles have actions of eversion of the foot.

In contrast to the upper limb muscles, those in the lower limb tend to be much larger and are involved in powerful contractions for walking or running as well as isometric contractions in standing. As compared to the forearm, the leg has fewer muscles because the foot is not as involved in actions that require precision movements as the hand.

As you demonstrate the actions of muscles in the lower limb, refer to standard anatomical regions of the limb. The *thigh* refers to hip to knee; the *leg* refers to knee to ankle; and the *foot* refers to ankle, metatarsal area, and toes. Several muscles in the hip and lower limb cross more than one joint and involve movements of both joints. Examples of muscles crossing more than one joint include the sartorius and gastrocnemius.

Procedure—Muscles of the Hip and Lower Limb

1. Study figures 13.1, 13.2, 13.3, and 13.4, and table 13.1.
2. Locate the following muscles in the human torso model and in the lower limb models. Also locate as many of them as possible in your body.

    muscles that move the thigh/hip joint
    - anterior hip muscles
        - iliopsoas group
            psoas major
            iliacus
    - posterior and lateral hip muscles
        - gluteus maximus
        - gluteus medius
        - gluteus minimus
        - tensor fasciae latae
        - piriformis

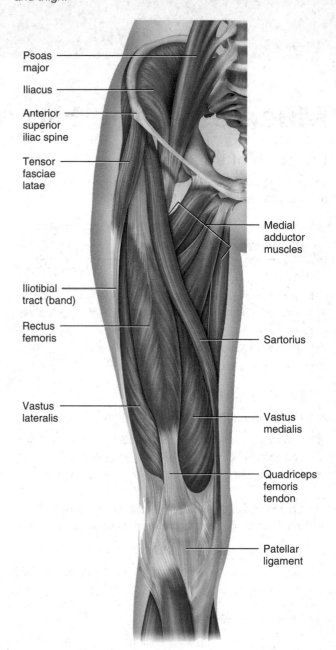

FIGURE 13.1 Muscles of the anterior right hip and thigh.

Psoas major
Iliacus
Anterior superior iliac spine
Tensor fasciae latae
Iliotibial tract (band)
Rectus femoris
Vastus lateralis
Medial adductor muscles
Sartorius
Vastus medialis
Quadriceps femoris tendon
Patellar ligament

- medial adductor muscles
    - pectineus
    - adductor longus
    - adductor magnus
    - adductor brevis
    - gracilis
3. Demonstrate the actions of these muscles in your body.

FIGURE 13.2 Selected individual muscles of the anterior right hip and medial thigh.

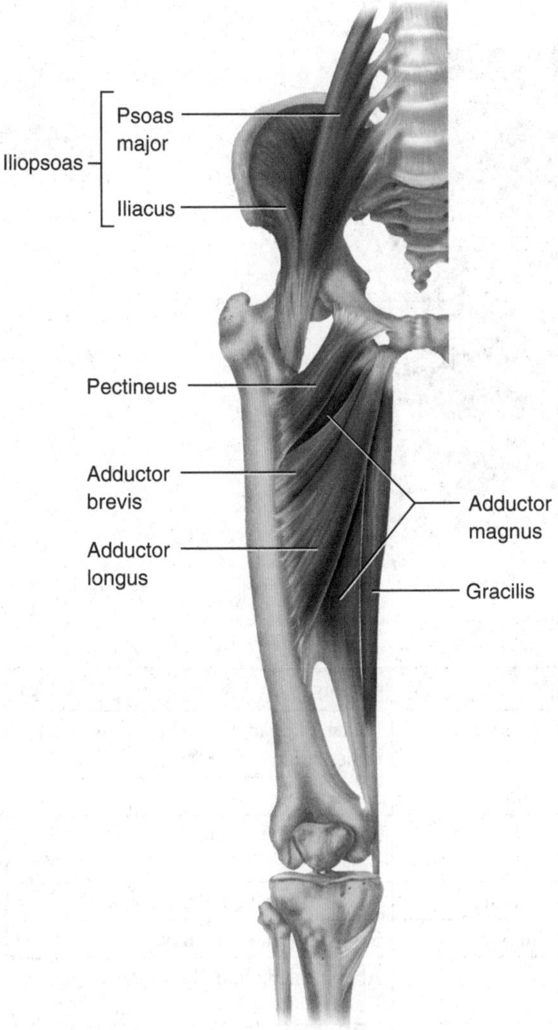

FIGURE 13.3 Muscles of the posterior right hip and thigh.

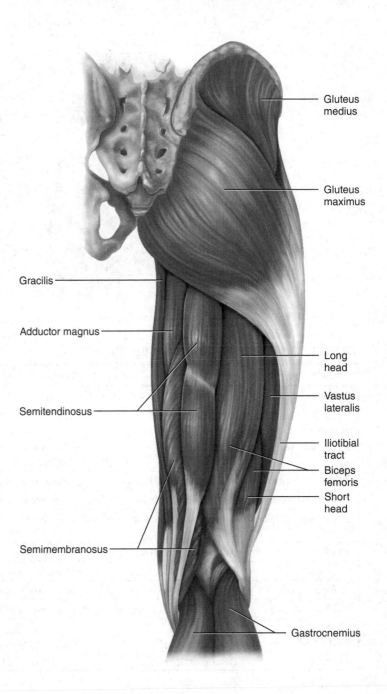

Iliopsoas — Psoas major

Iliacus

Pectineus

Adductor brevis

Adductor longus

Adductor magnus

Gracilis

Gluteus medius

Gluteus maximus

Gracilis

Adductor magnus

Semitendinosus

Long head

Vastus lateralis

Iliotibial tract

Biceps femoris

Short head

Semimembranosus

Gastrocnemius

FIGURE 13.4 Superficial and deep posterior hip muscles.

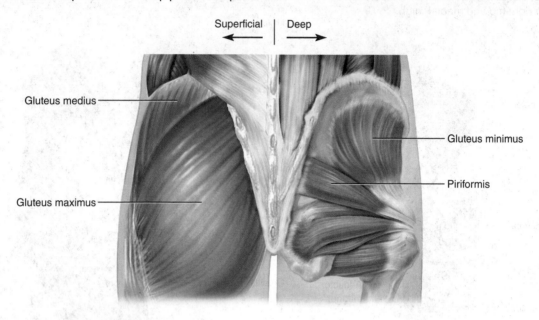

Superficial ← | Deep →

Gluteus medius

Gluteus minimus

Piriformis

Gluteus maximus

TABLE 13.1 Muscles That Move the Thigh

Muscle	Origin	Insertion	Action
Psoas major	Intervertebral discs, bodies, and transverse processes of T12–L5	Lesser trochanter of femur	Flexes thigh at hip; flexes trunk at hip (when thigh is fixed)
Iliacus	Iliac fossa of ilium	Lesser trochanter of femur	Same as psoas major
Gluteus maximus	Sacrum, coccyx, and posterior iliac crest	Gluteal tuberosity of femur and iliotibial tract	Extends thigh at hip
Gluteus medius	Lateral surface of ilium	Greater trochanter of femur	Abducts and medially rotates thigh
Gluteus minimus	Lateral surface of ilium	Greater trochanter of femur	Same as gluteus medius
Tensor fasciae latae	Iliac crest and anterior superior iliac spine	Iliotibial tract	Abducts and medially rotates thigh
Piriformis	Anterior surface of sacrum	Greater trochanter of femur	Abducts and laterally rotates thigh
Pectineus	Pubis	Femur distal to lesser trochanter	Adducts and flexes thigh
Adductor longus	Pubis near pubic symphysis	Linea aspera of femur	Adducts and flexes thigh
Adductor magnus	Pubis and ischial tuberosity	Linea aspera of femur	Adducts thigh; posterior portion extends thigh, and anterior portion flexes thigh
Adductor brevis	Pubis	Linea aspera of femur	Adducts thigh
Gracilis	Pubis	Medial surface of tibia	Adducts thigh and flexes leg at knee

4. Locate the origins and insertions of these muscles in the human skeleton.
5. Study figures 13.1, 13.3, and table 13.2.
6. Locate the following muscles in the human torso model and in the lower limb models. Also locate as many of them as possible in your body.

    muscles that move the leg/knee joint
    - anterior thigh muscles
        - sartorius
        - quadriceps femoris group
            rectus femoris
            vastus lateralis
            vastus medialis
            vastus intermedius
    - posterior thigh muscles
        - hamstring group
            biceps femoris
            semitendinosus
            semimembranosus

7. Demonstrate the actions of these muscles in your body.
8. Locate the origins and insertions of these muscles in the human skeleton.
9. Complete Part A of Laboratory Assessment 13.
10. Study figures 13.5, 13.6, and 13.7, and table 13.3.

TABLE 13.2 Muscles That Move the Leg

Muscle	Origin	Insertion	Action
Sartorius	Anterior superior iliac spine	Proximal medial surface of tibia	Flexes, abducts, and laterally rotates thigh at hip; flexes leg at knee
Quadriceps Femoris Group			
Rectus femoris	Anterior inferior iliac spine and superior margin of acetabulum	Patella by common quadriceps tendon, which continues as patellar ligament to tibial tuberosity	Extends leg at knee; flexes thigh at hip
Vastus lateralis	Greater trochanter and linea aspera of femur		Extends leg at knee
Vastus medialis	Linea aspera of femur		Extends leg at knee
Vastus intermedius	Anterior and lateral surfaces of femur		Extends leg at knee
Hamstring Group			
Biceps femoris	Ischial tuberosity (long head) and linea aspera of femur (short head)	Head of fibula and lateral condyle of tibia	Flexes leg at knee; rotates leg laterally; extends thigh
Semitendinosus	Ischial tuberosity	Proximal medial surface of tibia	Flexes leg at knee; rotates leg medially; extends thigh
Semimembranosus	Ischial tuberosity	Posterior medial condyle of tibia	Flexes leg at knee; rotates leg medially; extends thigh

FIGURE 13.5 Muscles of the anterior right leg (a and b).

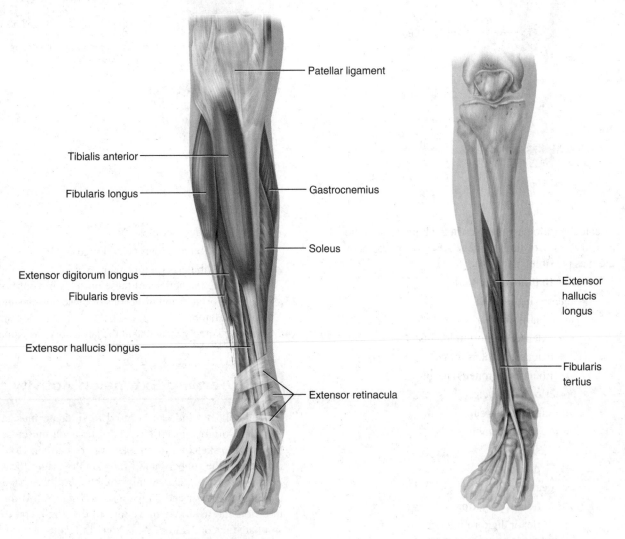

(a) Superficial view

(b) Selected individual muscles

123

FIGURE 13.6 Muscles of the posterior right leg (a and b).

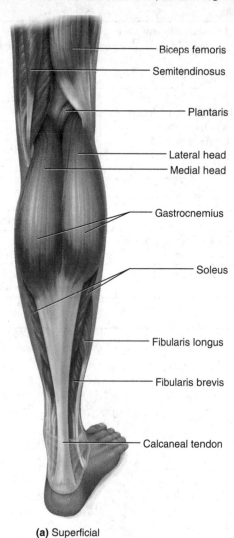

Biceps femoris
Semitendinosus
Plantaris
Lateral head
Medial head
Gastrocnemius
Soleus
Fibularis longus
Fibularis brevis
Calcaneal tendon

(a) Superficial

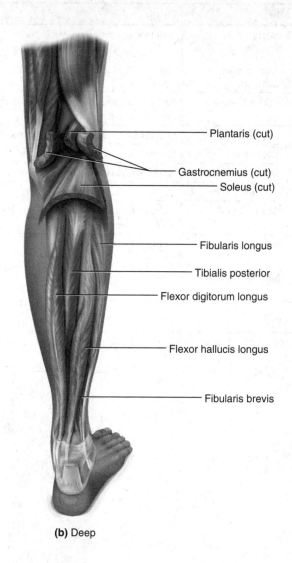

Plantaris (cut)
Gastrocnemius (cut)
Soleus (cut)
Fibularis longus
Tibialis posterior
Flexor digitorum longus
Flexor hallucis longus
Fibularis brevis

(b) Deep

11. Locate the following muscles in the human torso model and in the lower limb models. Also locate as many of them as possible in your body.

   muscles that move the foot
   - anterior leg muscles
     - tibialis anterior
     - extensor digitorum longus
     - extensor hallucis longus
     - fibularis (peroneus) tertius
   - posterior leg muscles
     - superficial group
       gastrocnemius
       soleus
       plantaris
     - deep group
       tibialis posterior
       flexor digitorum longus
       flexor hallucis longus

   - lateral leg muscles
     - fibularis (peroneus) longus
     - fibularis (peroneus) brevis
12. Demonstrate the actions of these muscles in your body.
13. Locate the origins and insertions of these muscles in the human skeleton.
14. Complete Parts B, C, and D of the laboratory assessment.

Learning Extension Activity

A long rubber band can be used to simulate muscle locations, origins, insertions, and actions on muscular models, the skeleton, or on a laboratory partner. Hold one end of the rubber band firmly on the origin location of a muscle, then slightly stretch the rubber band and hold the other end on the insertion site. Allow the insertion end to slowly move toward the origin end to simulate the contraction and action of the muscle.

FIGURE 13.7 Muscles of the lateral right leg.

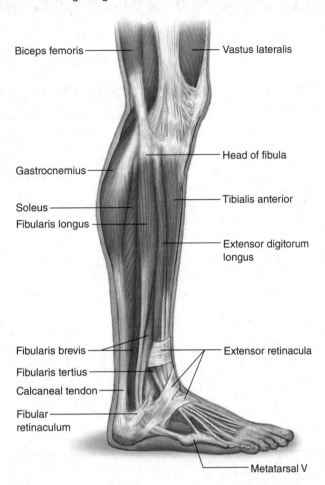

Biceps femoris — — Vastus lateralis

Gastrocnemius — — Head of fibula

Soleus — — Tibialis anterior

Fibularis longus — — Extensor digitorum longus

Fibularis brevis — — Extensor retinacula

Fibularis tertius —

Calcaneal tendon —

Fibular retinaculum —

— Metatarsal V

TABLE 13.3 Muscles That Move the Foot

Muscle	Origin	Insertion	Action
Tibialis anterior	Lateral condyle and proximal tibia	Medial cuneiform and metatarsal I	Dorsiflexion and inversion of foot
Extensor digitorum longus	Lateral condyle of tibia and anterior surface of fibula	Dorsal surfaces of second and third phalanges of toes 2–5	Extends toes 2–5 and dorsiflexion
Extensor hallucis longus	Anterior surface of fibula	Distal phalanx of the great toe	Extends great toe and dorsiflexion
Fibularis tertius	Anterior distal surface of fibula	Dorsal surface of metatarsal V	Dorsiflexion and eversion of foot
Gastrocnemius	Lateral and medial condyles of femur	Calcaneus via calcaneal tendon	Plantar flexion of foot and flexes knee
Soleus	Head and shaft of fibula and posterior surface of tibia	Calcaneus via calcaneal tendon	Plantar flexion of foot
Plantaris	Superior to lateral condyle of femur	Calcaneus	Plantar flexion of foot and flexes knee
Tibialis posterior	Posterior tibia and fibula	Tarsals (several) and metatarsals II–IV	Plantar flexion and inversion of foot
Flexor digitorum longus	Posterior surface of tibia	Distal phalanges of toes 2–5	Flexes toes 2–5 and plantar flexion and inversion of foot
Flexor hallucis longus	Posterior distal fibula	Distal phalanx of great toe	Flexes great toe and plantar flexes foot
Fibularis longus	Head and shaft of fibula and lateral condyle of tibia	Medial cuneiform and metatarsal I	Plantar flexion and eversion of foot; also supports arch
Fibularis brevis	Distal fibula	Metatarsal V	Plantar flexion and eversion of foot

Name _____

Date _____

Section _____

The Ⓐ corresponds to the indicated outcome(s) found at the beginning of the laboratory exercise.

Muscles of the Hip and Lower Limb

Part A   Assessments

Identify the muscles indicated in the hip and thigh in figures 13.8 and 13.9.

FIGURE 13.8   Label the right anterior muscles of the hip and thigh. Ⓐ

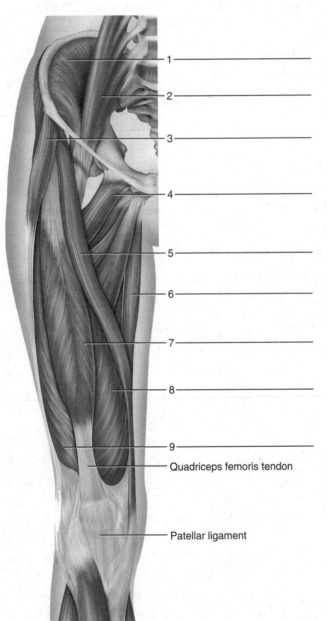

1 _____

2 _____

3 _____

4 _____

5 _____

6 _____

7 _____

8 _____

9 _____

—— Quadriceps femoris tendon

—— Patellar ligament

FIGURE 13.9   Label the right posterior hip and thigh muscles of a cadaver, using the terms provided. The gluteus maximus has been removed. Ⓐ

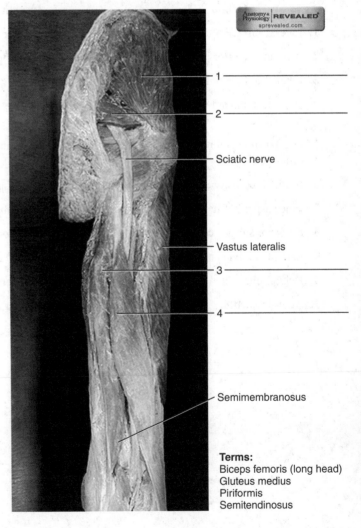

1 _____

2 _____

—— Sciatic nerve

—— Vastus lateralis

3 _____

4 _____

—— Semimembranosus

Terms:
Biceps femoris (long head)
Gluteus medius
Piriformis
Semitendinosus

Part B Assessments

Match the muscles in column A with the actions in column B. Place the letter of your choice in the space provided. 🄰

Column A	Column B
a. Biceps femoris	_____ **1.** Adducts thigh and flexes knee
b. Fibularis (peroneus) longus	_____ **2.** Plantar flexion and eversion of foot
c. Gluteus medius	_____ **3.** Flexes thigh at the hip
d. Gracilis	_____ **4.** Abducts and flexes thigh and rotates it laterally; flexes leg at knee
e. Psoas major and iliacus	_____ **5.** Abducts thigh and rotates it medially
f. Quadriceps femoris group	_____ **6.** Plantar flexion and inversion of foot
g. Sartorius	_____ **7.** Flexes leg at the knee and laterally rotates leg
h. Tibialis anterior	_____ **8.** Extends leg at the knee
i. Tibialis posterior	_____ **9.** Dorsiflexion and inversion of foot

Part C Assessments

Name the muscle indicated by the following combinations of origin and insertion. 🄰

Origin	Insertion	Muscle
1. Lateral surface of ilium	Greater trochanter of femur	_____
2. Anterior superior iliac spine	Medial surface of proximal tibia	_____
3. Lateral and medial condyles of femur	Calcaneus	_____
4. Iliac crest and anterior superior iliac spine	Iliotibial tract	_____
5. Greater trochanter and linea aspera of femur	Patella to tibial tuberosity	_____
6. Ischial tuberosity	Medial surface of proximal tibia	_____
7. Linea aspera of femur	Patella to tibial tuberosity	_____
8. Posterior surface of tibia	Distal phalanges of four lateral toes	_____
9. Lateral condyle and proximal tibia	Medial cuneiform and first metatarsal	_____
10. Ischial tuberosity and pubis	Linea aspera of femur	_____
11. Anterior sacrum	Greater trochanter of femur	_____

Part D Assessments

Identify the muscles indicated in figure 13.10.

FIGURE 13.10 Label these muscles that appear as lower limb surface features in these photographs (a and b), by placing the correct numbers in the spaces provided. ⚠

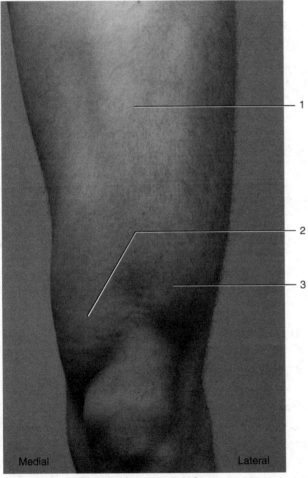

(a) Left thigh, anterior view

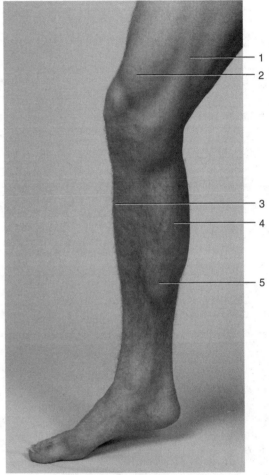

(b) Right lower limb, medial view

_____ Rectus femoris
_____ Vastus lateralis
_____ Vastus medialis

_____ Gastrocnemius
_____ Sartorius
_____ Soleus
_____ Tibialis anterior
_____ Vastus medialis

129

NOTES

Nervous Tissue and Nerves

Purpose of the Exercise

To review the characteristics of nervous tissue and to observe neurons, neuroglia, and various features of the nerves.

Materials Needed

Compound light microscope
Prepared microscope slides of the following:
Spinal cord (smear)
Dorsal root ganglion (section)
Neuroglia (astrocytes)
Peripheral nerve (cross section and longitudinal section)
Neuron model

For Learning Extension Activity:
Prepared microscope slide of Purkinje cells from cerebellum

Learning Outcomes

After completing this exercise, you should be able to

1. Describe and locate the characteristics of nervous tissue.
2. Distinguish structural and functional characteristics between neurons and neuroglia.
3. Identify and sketch the major structures of a neuron and a nerve.

Pre-Lab

Carefully read the introductory material and examine the entire lab. Be familiar with the structures and functions of nervous tissue and nerves from lecture or the textbook. Answer the pre-lab questions.

Pre-Lab Questions: Select the correct answer for each of the following questions:

1. The cell body of a neuron contains the
   a. nucleus. **b.** dendrites.
   c. axon. **d.** neuroglia.
2. A multipolar neuron contains
   a. one dendrite and many axons.
   b. many dendrites and one axon.
   c. one dendrite and one axon.
   d. a single process with the dendrite and axon.
3. Neuroglia that produce myelin insulation in the CNS are
   a. microglia. **b.** astrocytes.
   c. ependymal cells. **d.** oligodendrocytes.
4. The PNS contains
   a. 12 pairs of cranial nerves only.
   b. 31 pairs of spinal nerves only.
   c. 12 pairs of cranial nerves and 31 pairs of spinal nerves.
   d. 43 pairs of spinal nerves.
5. Schwann cells
   a. are only in the brain.
   b. are only in the spinal cord.
   c. are throughout the CNS.
   d. have a myelin sheath and neurilemma.
6. A _____ neuron is the most common structural neuron in the brain and spinal cord.
   a. multipolar **b.** tripolar
   c. bipolar **d.** unipolar
7. Sensory neurons conduct impulses from the spinal cord to a muscle or a gland.
   True _____ False _____
8. Astrocytes have contacts between blood vessels and neurons in the CNS.
   True _____ False _____

Nervous tissue, which occurs in the brain, spinal cord, and peripheral nerves, contains *neurons* (nerve cells) and *neuroglia* (neuroglial cells; glial cells). Neurons are irritable (excitable) and easily respond to stimuli, resulting in the conduction of an action potential (nerve impulse). A neuron contains a cell body with a nucleus and most of the cytoplasm, and elongated cell processes (dendrites and axons) along which impulse conductions occur. Neurons can be classified according to structural variations of their cell processes: multipolar with many dendrites and one axon (nerve fiber), bipolar with one dendrite and one axon, and unipolar with a single process where the dendrite leads directly into the axon. Functional classifications are used according to the direction of the impulse conduction: sensory (afferent) neurons conduct impulses from receptors to the central nervous system (CNS), interneurons (association neurons) conduct impulses within the CNS, and motor (efferent) neurons conduct impulses away from the CNS to effectors (muscles or glands).

Neuroglia (supportive cells) are located in close association with neurons. Four types of neuroglia are in the CNS: astrocytes, microglia, oligodendrocytes, and ependymal cells. Astrocytes are numerous, and their branches support neurons and blood vessels. Microglia can phagocytize microorganisms and nerve tissue debris. Oligodendrocytes produce the myelin insulation in the CNS. Ependymal cells line the brain ventricles and secrete CSF (cerebrospinal fluid), and the cilia on their apical surfaces aid the circulation of the CSF.

Two types of neuroglia are located in the peripheral nervous system (PNS): Schwann cells and satellite cells. Schwann cells surround nerve fibers numerous times. The wrappings nearest the nerve fiber represent the myelin sheath and serve to insulate and increase the impulse speed, while the last wrapping, called the neurilemma, contains most of the cytoplasm and the nucleus and functions in nerve fiber regeneration in PNS neurons. Satellite cells surround and support the cell body regions, called ganglia, of peripheral neurons.

The PNS contains 12 pairs of cranial nerves and 31 pairs of spinal nerves, all containing parallel axons of neurons and neuroglia representing the nervous tissue components. Most nerves are mixed in that they contain both sensory and motor neurons, but some contain only sensory or motor components. The nerves represent an organ structure as they also contain small blood vessels and fibrous connective tissue. The fibrous connective tissue around each nerve fiber and the Schwann cell is the endoneurium; a bundle of nerve fibers, representing a fascicle, is surrounded by a perineurium; and the entire nerve is surrounded by an epineurium. The fibrous connective components of a nerve provide protection and stretching during body movements.

Procedure A—Nervous Tissue

In this procedure you will concentrate on more detail of nervous tissue than was observed when nervous tissue was first examined in the earlier tissue lab. Two tables of nervous tissue include the structural and functional characteristics of neurons, and neuroglia within the central nervous system (CNS) and the peripheral nervous system (PNS). Diagrams of neurons and neuroglia are provided as sources of comparison for your microscopic observations using the compound microscope. When you make sketches in the laboratory assessment, include all the observed structures with labels.

1. Examine tables 14.1 and 14.2 and figures 14.1, 14.2, and 14.3.
2. Complete Parts A and B of Laboratory Assessment 14.
3. Obtain a prepared microscope slide of a spinal cord smear. Using low-power magnification, search the slide

TABLE 14.1 Structural and Functional Types of Neurons

Classified by Structure		
Type	**Structural Characteristics**	**Location**
Multipolar neuron	Cell body with one axon and multiple dendrites	Most common type of neuron in the brain and spinal cord
Bipolar neuron	Cell body with one axon and one dendrite	In receptor parts of the eyes, nose, and ears; rare type
Unipolar neuron	Cell body with a single process that divides into two branches and functions as an axon; only the receptor ends of the peripheral (distal) process function as dendrites	Most sensory neurons
Classified by Function		
Type	**Functional Characteristics**	**Structural Characteristics**
Sensory (afferent) neuron	Conducts impulses from receptors in peripheral body parts into the brain or spinal cord	Most unipolar; some bipolar
Interneuron	Transmits impulses between neurons in the brain and spinal cord	Multipolar
Motor (efferent) neuron	Conducts impulses from the brain or spinal cord out to muscles or glands (effectors)	Multipolar

TABLE 14.2   Types of Neuroglia in the CNS and PNS

Type	Structural Characteristics	Functions
CNS (brain and spinal cord)		
Astrocytes	Star-shaped cells contacting neurons and blood vessels; most abundant type	Structural support; formation of scar tissue to replace damaged neurons; transport of substances between blood vessels and neurons; communicate with one another and with neurons; remove excess ions and neurotransmitters
Microglia	Small cells with slender cellular processes	Structural support; phagocytosis of microorganisms and damaged tissue
Oligodendrocytes	Shaped like astrocytes, but with fewer cellular processes; processes wrap around axons	Form myelin sheaths in the brain and spinal cord
Ependyma	Cuboidal cells lining cavities of the brain and spinal cord	Secrete and assist in circulation of cerebrospinal fluid (CSF)
PNS (peripheral nerves)		
Schwann cells	Cells that wrap tightly around the axons of peripheral neurons	Speed neurotransmission
Satellite cells	Flattened cells that surround cell bodies of neurons in ganglia	Support ganglia

FIGURE 14.1   Structural types of neurons (a) multipolar neuron, (b) bipolar neuron, and (c) unipolar neuron.

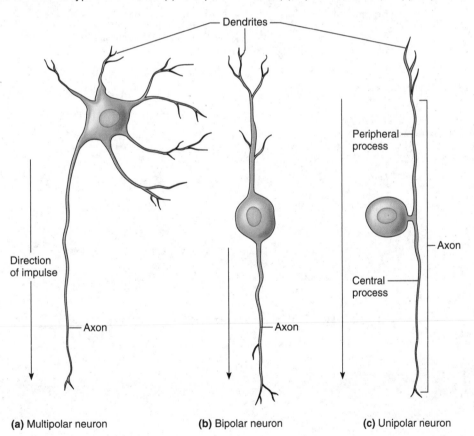

(a) Multipolar neuron   (b) Bipolar neuron   (c) Unipolar neuron

FIGURE 14.2 Diagram of a multipolar motor neuron.

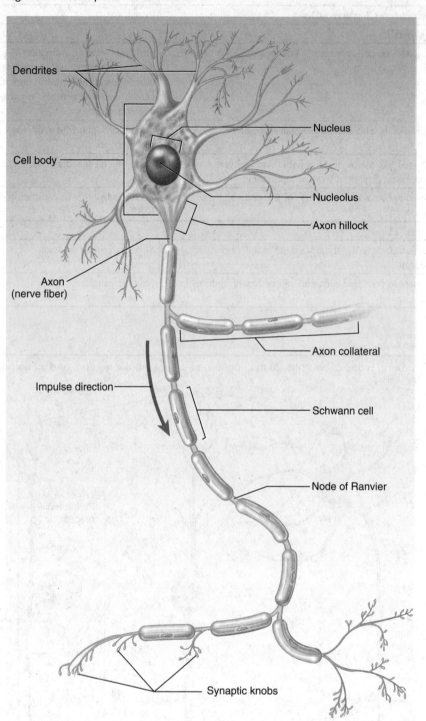

Dendrites

Nucleus

Cell body

Nucleolus

Axon hillock

Axon
(nerve fiber)

Axon collateral

Impulse direction

Schwann cell

Node of Ranvier

Synaptic knobs

FIGURE 14.3 Diagram of a cross section of a myelinated axon (nerve fiber) of a spinal nerve.

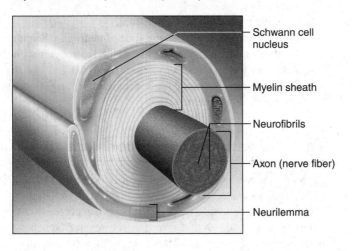

- Schwann cell nucleus
- Myelin sheath
- Neurofibrils
- Axon (nerve fiber)
- Neurilemma

and locate the relatively large, deeply stained cell bodies of multipolar motor neurons (fig. 14.4).

4. Observe a single multipolar motor neuron, using high-power magnification, and note the following features:

cell body (soma)
- nucleus
- nucleolus
- Nissl bodies (chromatophilic substance)— a type of rough ER in neurons
- neurofibrils—threadlike structures extending into axon
- axon hillock—origin region of axon

dendrites—conduct impulses toward cell body

axon (nerve fiber)—conducts impulse away from cell body

Compare the slide to the neuron model and to figure 14.4. Also note small, darkly stained nuclei of neuroglia around the motor neuron.

FIGURE 14.4 Micrograph of a multipolar neuron and neuroglia from a spinal cord smear (600×).

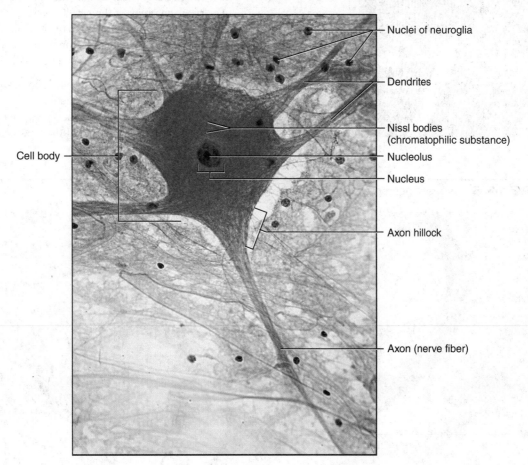

- Nuclei of neuroglia
- Dendrites
- Nissl bodies (chromatophilic substance)
- Nucleolus
- Nucleus
- Axon hillock
- Cell body
- Axon (nerve fiber)

135

5. Sketch and label a motor (efferent) neuron in the space provided in Part C of the laboratory assessment.
6. Obtain a prepared microscope slide of a dorsal root ganglion. Search the slide and locate a cluster of sensory neuron cell bodies. You also may note bundles of nerve fibers passing among groups of neuron cell bodies (fig. 14.5).

7. Sketch and label a sensory (afferent) neuron cell body in the space provided in Part C of the laboratory assessment.
8. Examine figure 14.6 and table 14.2.
9. Obtain a prepared microscope slide of neuroglia. Search the slide and locate some darkly stained astrocytes with numerous long, slender processes (fig. 14.7).
10. Sketch a neuroglia in the space provided in Part C of the laboratory assessment.

FIGURE 14.5 Micrograph of a dorsal root ganglion (100×).

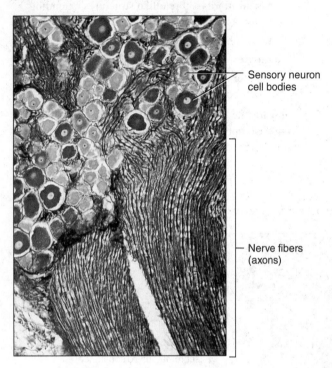

Sensory neuron cell bodies

Nerve fibers (axons)

Learning Extension Activity

Obtain a prepared microscope slide of Purkinje cells (fig. 14.8). To locate these neurons, search the slide for large, flask-shaped cell bodies. Each cell body has one or two large, thick dendrites that give rise to extensive branching networks of dendrites. These large cells are located in a particular region of the brain (cerebellar cortex).

Procedure B—Nerves

In this procedure you will compare an illustration of a peripheral spinal nerve with actual microscopic views of nerves. A nerve is an organ of the nervous system and contains various tissues in combination with the nervous tissue. The nerve contains the nerve fibers (axons) of the neurons and the Schwann cells representing some neuroglia in the PNS.

1. Study figure 14.9 of a cross section of a spinal nerve. Examine the fibrous connective tissue pattern including the epineurium around the entire nerve, the perineurium around fascicles, and the endoneurium around an

FIGURE 14.6 Types of neuroglia in the CNS.

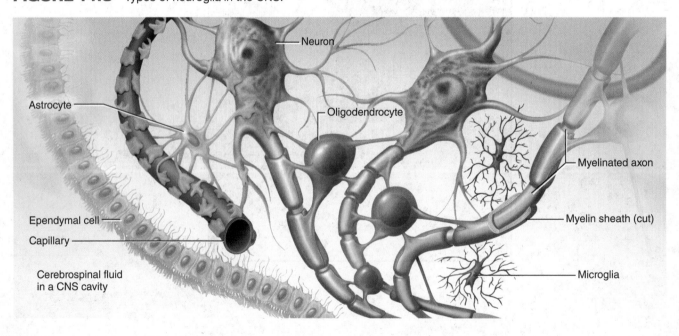

Neuron

Astrocyte

Oligodendrocyte

Myelinated axon

Ependymal cell

Capillary

Myelin sheath (cut)

Cerebrospinal fluid in a CNS cavity

Microglia

FIGURE 14.7 Micrograph of astrocytes (1,000×).

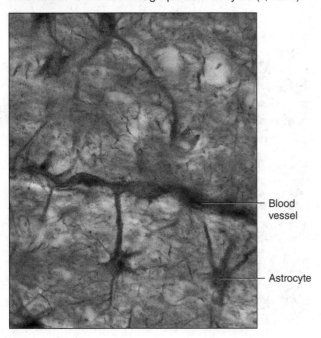

Blood vessel

Astrocyte

FIGURE 14.8 Large multipolar Purkinje cell from cerebellum of the brain (400×).

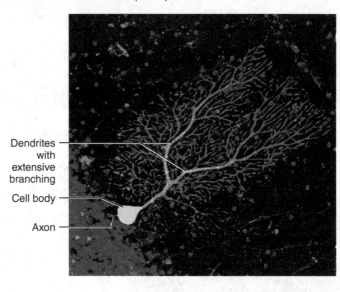

Dendrites with extensive branching

Cell body

Axon

FIGURE 14.9 Diagram of a cross section of a peripheral spinal nerve.

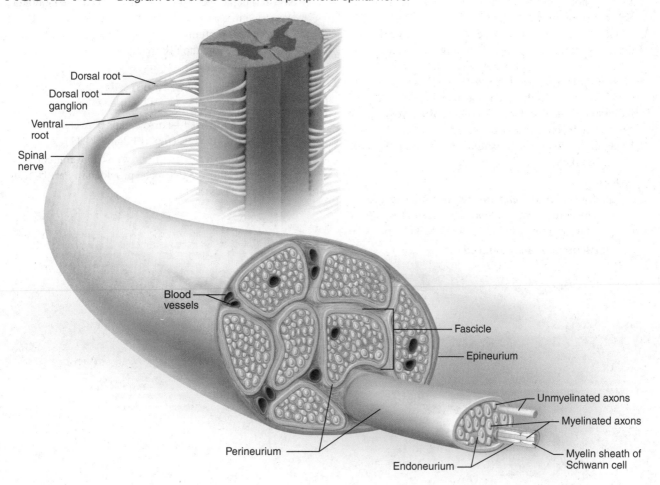

Dorsal root

Dorsal root ganglion

Ventral root

Spinal nerve

Blood vessels

Fascicle

Epineurium

Unmyelinated axons

Myelinated axons

Myelin sheath of Schwann cell

Perineurium

Endoneurium

137

FIGURE 14.10 Cross section of a bundle of neurons within a nerve (400×).

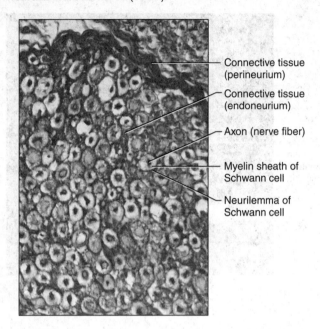

- Connective tissue (perineurium)
- Connective tissue (endoneurium)
- Axon (nerve fiber)
- Myelin sheath of Schwann cell
- Neurilemma of Schwann cell

FIGURE 14.11 Longitudinal section of a nerve (2,000×).

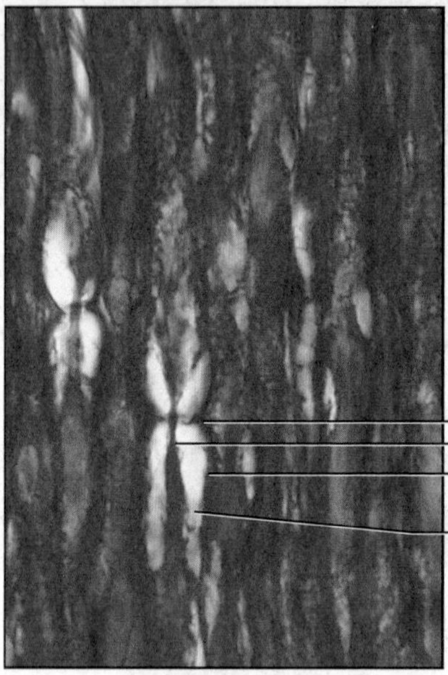

- Node of Ranvier
- Axon (nerve fiber)
- Neurilemma of Schwann cell
- Myelin sheath of Schwann cell

individual axon and Schwann cell. Reexamine figure 14.3 to note additional cross-sectional detail of the Schwann cell layers.

2. Obtain a prepared microscope slide of a nerve. Locate the cross section of the nerve and note the many round nerve fibers inside. Also note the dense layer of connective tissue (perineurium) that encircles a fascicle of nerve fibers and holds them together in a bundle. The individual nerve fibers are surrounded by a layer of more delicate connective tissue (endoneurium) (fig. 14.10).

3. Using high-power magnification, observe a single nerve fiber and note the following features:

   axon

   myelin sheath of Schwann cell around the axon (most of the myelin may have been dissolved and lost during the slide preparation)

   neurilemma of Schwann cell

4. Sketch and label a nerve fiber with Schwann cell (cross section) in the space provided in Part D of the laboratory assessment.

5. Locate the longitudinal section of the nerve on the slide (fig. 14.11). Note the following:

   axons

   myelin sheath of Schwann cells

   neurilemma of Schwann cells

   nodes of Ranvier—narrow gaps between the Schwann cells (figs. 14.2 and 14.11)

6. Sketch and label a nerve fiber with Schwann cell (longitudinal section) in the space provided in Part D of the laboratory assessment.

Name _____

Date _____

Section _____

The 🅐 corresponds to the indicated outcome(s) found at the beginning of the
laboratory exercise.

Nervous Tissue and Nerves

Part A   Assessments

Match the terms in column A with the descriptions in column B. Place the letter of your choice in the space provided. 🅐 🅐

Column A

a. Astrocyte
b. Axon
c. Collateral
d. Dendrite
e. Myelin
f. Neurilemma
g. Neurofibrils
h. Nissl bodies (chromatophilic substance)
i. Unipolar neuron

Column B

_____ 1. Sheath of Schwann cell containing cytoplasm and nucleus that encloses myelin

_____ 2. Corresponds to rough endoplasmic reticulum in other cells

_____ 3. Network of threadlike structures within cell body and extending into axon

_____ 4. Substance of Schwann cell composed of lipoprotein that insulates axons and increases impulse speed

_____ 5. Neuron process with many branches that conducts an action potential (impulse) toward the cell body

_____ 6. Branch of an axon

_____ 7. Star-shaped neuroglia between neurons and blood vessels

_____ 8. Nerve fiber arising from a slight elevation of the cell body that conducts an action potential (impulse) away from the cell body

_____ 9. Possesses a single process from the cell body

Part B   Assessments

Match the terms in column A with the descriptions in column B. Place the letter of your choice in the space provided. 🅐 🅐

Column A

a. Effector
b. Ependymal cell
c. Ganglion
d. Interneuron (association neuron)
e. Microglia
f. Motor (efferent) neuron
g. Oligodendrocyte
h. Sensory (afferent) neuron

Column B

_____ 1. Transmits impulse from sensory to motor neuron within central nervous system

_____ 2. Transmits impulse out of the brain or spinal cord to effectors (muscles and glands)

_____ 3. Transmits impulse into brain or spinal cord from receptors

_____ 4. Myelin-forming neuroglia in brain and spinal cord

_____ 5. Phagocytic neuroglia

_____ 6. Structure capable of responding to motor impulse

_____ 7. Specialized mass of neuron cell bodies outside the brain or spinal cord

_____ 8. Cells that line cavities of the brain and secrete cerebrospinal fluid

Part C Assessments

In the space that follows, sketch the indicated cells. Label any of the cellular structures observed, and indicate the magnification of each sketch. 🄰 🄰

Motor neuron (_____×)

Sensory neuron cell body (_____×)

Neuroglia (_____×)

Part D Assessments

In the space that follows, sketch the indicated view of a nerve fiber (axon). Label any structures observed, and indicate the magnification of each sketch. 🄰 🄰

Nerve fiber cross section with Schwann cell (_____×)

Nerve fiber longitudinal section with Schwann cell (_____×)

Spinal Cord, Spinal Nerves, and Meninges

Purpose of the Exercise

To review the characteristics of the spinal cord, spinal nerves, and meninges and to observe the major features of these structures.

Materials Needed

Compound light microscope
Prepared microscope slide of a spinal cord cross section with spinal nerve roots
Spinal cord model with meninges
Vertebral column model with spinal nerves

For Demonstration Activity:
Preserved spinal cord with meninges intact

Learning Outcomes

After completing this exercise, you should be able to

1. Identify the major features and functions of the spinal cord.

2. Locate the distribution and features of the spinal nerves.

3. Arrange the layers of the meninges and describe the structure of each.

Pre-Lab

Carefully read the introductory material and examine the entire lab. Be familiar with the structures and functions of the spinal cord, spinal nerves, and meninges from lecture or the textbook. Answer the pre-lab questions.

Pre-Lab Questions: Select the correct answer for each of the following questions:

1. The inferior end of the adult spinal cord ends
   a. inferior to L1.   **b.** inferior to L5.
   c. in the sacrum.   **d.** in the coccyx.

2. The dorsal root of spinal nerves contains
   a. interneurons.   **b.** sensory neurons.
   c. motor neurons.   **d.** sensory and motor neurons.

3. The _____ is the most superficial membrane of the meninges.
   a. subarachnoid space   **b.** pia mater
   c. arachnoid mater   **d.** dura mater

4. Which of the following is *not* part of the gray matter of the spinal cord?
   a. gray commissure   **b.** posterior horn
   c. lateral funiculus   **d.** lateral horn

5. The central canal of the spinal cord is located within the
   a. white matter.   **b.** epidural space.
   c. gray commissure.   **d.** subarachnoid space.

6. The brachial plexus is formed from components of spinal nerves C5–T1.
   True _____   False _____

7. There are eight pairs of cervical spinal nerves.
   True _____   False _____

8. The major ascending (sensory) and descending (motor) tracts compose the gray matter of the spinal cord.
   True _____   False _____

The spinal cord is a column of nerve fibers that extends down through the vertebral canal. Together with the brain, it makes up the central nervous system. In the cervical and lumbar regions of the spinal cord, enlargements give rise to spinal nerves to the upper limbs and lower limbs, respectively. The inferior end portion of the spinal cord, the conus medullaris, is located just inferior to lumbar vertebra L1 in the adult.

There are 31 pairs of spinal nerves attached to the spinal cord. At close proximity to the cord, the spinal nerve has two branches: a dorsal (posterior) root containing sensory neurons and dorsal root ganglion, and a ventral (anterior) root containing motor neurons. The spinal nerves emerge through the nearby intervertebral foramina; however, most lumbar and sacral nerves extend inferiorly through the vertebral canal as the cauda equina to emerge in their respective regions of the vertebral column.

The spinal cord has a central region, the *gray matter,* with paired posterior, lateral, and anterior horns connected by a gray commissure containing the central canal. The gray matter is a processing center for spinal reflexes and synaptic integration. The *white matter* of the spinal cord, represented by paired posterior, lateral, and anterior funiculi (columns), contains ascending (sensory) and descending (motor) tracts. The specific names of the spinal tracts often reflect their respective origins and the destinations of the fibers. At various levels of the spinal cord, many of the tracts cross over (decussate) to the opposite side of the spinal cord or the brainstem.

The meninges consist of three layers of fibrous connective tissue membranes located between the bones of the skull and vertebral column and the soft tissues of the central nervous system. The most superficial layer, the *dura mater,* is a tough membrane with an epidural space containing blood vessels, loose connective tissue, and adipose tissue between the membrane and the vertebrae. A weblike *arachnoid mater* adheres to the inside of the dura mater. The subarachnoid space, located beneath the arachnoid mater, contains cerebrospinal fluid (CSF) and serves as a protective cushion for the spinal cord and brain. The delicate innermost membrane, the *pia mater,* adheres to the surface of the spinal cord and brain. The denticulate ligaments, extending from the pia mater to the dura mater, anchor the spinal cord. An inferior extension of the pia mater, the filum terminale, anchors the spinal cord to the coccyx.

Procedure A—Structure of the Spinal Cord

1. Study figures 15.1, 15.2, and 15.3
2. Obtain a prepared microscope slide of a spinal cord cross section. Use the low power of the microscope to locate the following features:

   posterior median sulcus
   anterior median fissure
   central canal
   gray matter
   - gray commissure
   - posterior (dorsal) horn
   - lateral horn
   - anterior (ventral) horn
   white matter
   - posterior (dorsal) funiculus (column)
   - lateral funiculus (column)
   - anterior (ventral) funiculus (column)

FIGURE 15.1 Features of the spinal cord cross section and surrounding structures.

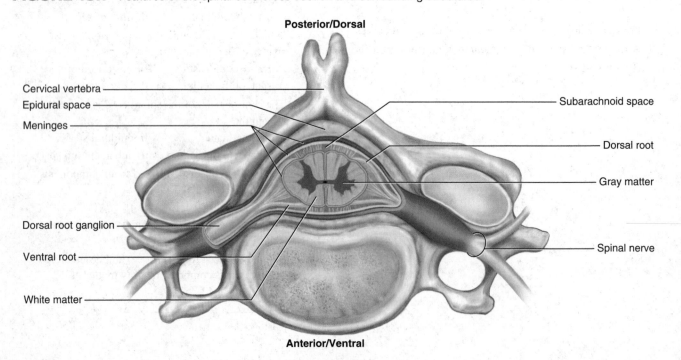

Posterior/Dorsal

Cervical vertebra
Epidural space
Meninges
Dorsal root ganglion
Ventral root
White matter

Subarachnoid space
Dorsal root
Gray matter
Spinal nerve

Anterior/Ventral

FIGURE 15.2 Cross section of the spinal cord, including the features of the white and gray matter.

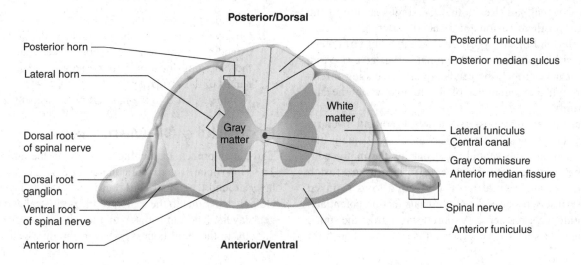

FIGURE 15.3 Cross section of a spinal cord with the three funiculi of the white matter shown on one side. Each funiculus contains specific tracts. Major ascending (sensory) and descending (motor) tracts (pathways) are shown only on one side of the spinal cord, but are located on both sides. Ascending tracts are in pink; descending tracts are in rust. This pattern varies with the level of the spinal cord. This pattern is representative of the midcervical region. (*Note:* These tracts are not visible as individually stained structures on microscope slides.)

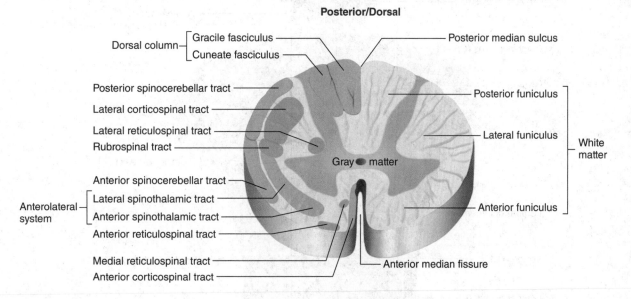

roots of spinal nerve
- dorsal roots
- dorsal root ganglia
- ventral roots

3. Observe the model of the spinal cord, and locate the features listed in step 2.
4. Complete Part A of Laboratory Assessment 15.

Procedure B—Spinal Nerves

There are 31 pairs of spinal nerves that emerge between bones of the vertebral column. Cervical spinal nerve pair C1 emerges between the occipital bone of the skull and the atlas. The other cervical nerves emerge inferior to the 7 cervical vertebrae, resulting in 8 pairs of cervical nerves. The rest of the spinal nerves emerge inferiorly to the corresponding vertebrae, resulting in 12 pairs of thoracic nerves, 5 pairs of lumbar nerves, 5 pairs of sacral nerves, and 1 coccygeal pair. Recall that the single sacrum of the adult is a result of fusions of 5 sacral vertebrae. Because the inferior end of the spinal cord of the adult terminates just inferior to L1, many of the spinal nerves give rise to a bundle of nerve roots, the cauda equina, extending inferiorly through the remainder of the lumbar vertebrae and the sacrum.

Each spinal nerve possesses a dorsal root and a ventral root adjacent to the spinal cord. The dorsal root contains the sensory neurons; the ventral root contains the motor neurons. The resulting main spinal nerve, formed from a fusion of the dorsal and ventral roots, is referred to as a mixed nerve because sensory and motor signals (action potentials) exist within the same nerve. A short distance from the spinal cord, some of spinal nerves merge (anastomose) and form

a weblike plexus. Major anastomoses include the cervical, brachial, lumbar, and sacral nerve plexuses.

1. Examine figures 15.4 and 15.5.
2. Observe a vertebral column model with spinal nerves extending from the intervertebral foramina. Compare the emerging locations of the spinal nerves with the names of the vertebrae.
3. Complete Part B of the laboratory assessment.

Procedure C—Meninges

Three membranes, or meninges, surround the entire CNS. The superficial, tough dura mater is a single meningeal layer around the spinal cord, but it is a double layer in the cranial cavity due to the fusion of the periosteal and meningeal layers. The dura mater of the cranial cavity adheres directly to the skull bones; however, the dura mater of the

FIGURE 15.4 Posterior view of the origins and categories of the 31 pairs of spinal nerves on the right. The plexuses formed by various spinal nerves are illustrated on the left.

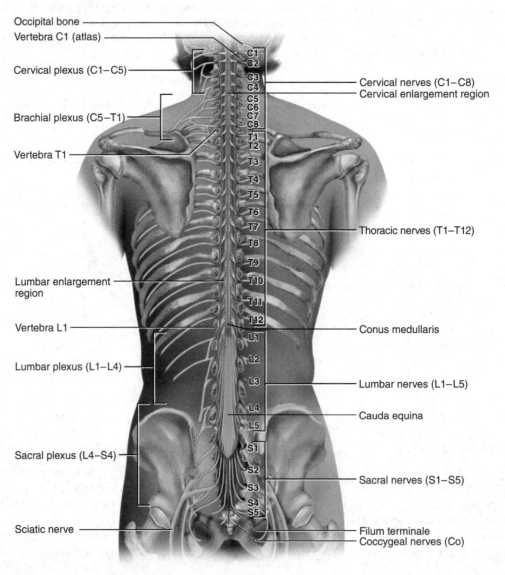

FIGURE 15.5 Posterior view of cervical region of spinal cord and associated nerves and meninges of a cadaver.

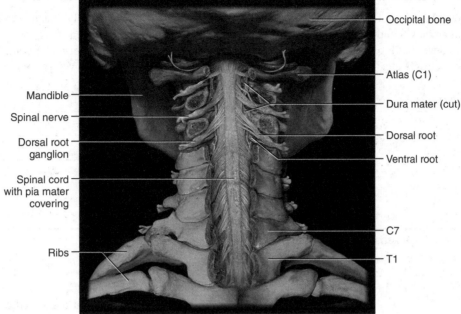

Occipital bone

Atlas (C1)

Dura mater (cut)

Dorsal root

Ventral root

C7

T1

Mandible

Spinal nerve

Dorsal root ganglion

Spinal cord with pia mater covering

Ribs

spinal cord has an epidural space between the membrane and the vertebrae. A subarachnoid space exists between the middle arachnoid mater and the pia mater that contains cerebrospinal fluid (CSF). The delicate pia mater closely adheres to the surface of the brain and spinal cord. Collectively, the meninges enclose and provide physical protective layers around the delicate brain and spinal cord. The CSF provides an additional protective cushion from sudden jolts to the head or the back.

1. Study figures 15.6 and 15.7. Review figures 15.1 and 15.5.
2. Observe the spinal cord model with meninges and locate the following features:

    dura mater (dural sheath)

    arachnoid mater (membrane)

    subarachnoid space

    denticulate ligament

    pia mater

3. Complete Part C of the laboratory assessment.

FIGURE 15.6 The meninges, associated near the spinal cord and spinal nerves.

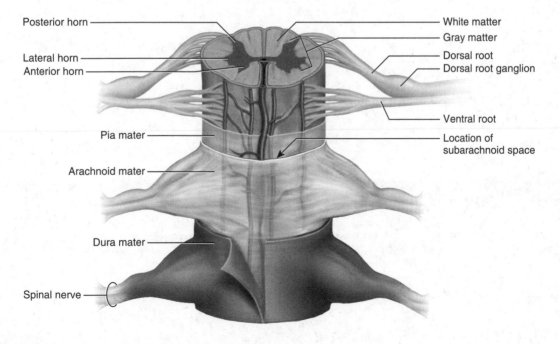

Posterior horn

Lateral horn

Anterior horn

Pia mater

Arachnoid mater

Dura mater

Spinal nerve

White matter

Gray matter

Dorsal root

Dorsal root ganglion

Ventral root

Location of subarachnoid space

Demonstration Activity

Observe the preserved section of spinal cord. Note the heavy covering of dura mater, firmly attached to the cord on each side by a set of ligaments (denticulate ligaments) originating in the pia mater. The intermediate layer of meninges, the arachnoid mater, is devoid of blood vessels, but in a live human being, the space beneath this layer contains cerebrospinal fluid. The pia mater, closely attached to the surface of the spinal cord, contains many blood vessels. What are the functions of these layers?

_____

_____

_____

FIGURE 15.7   Meninges associated with the brain.

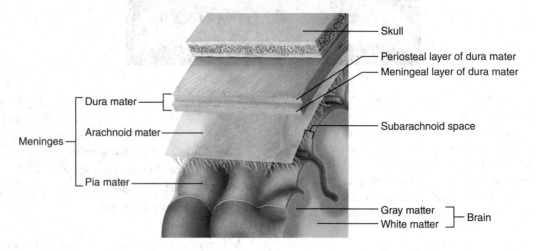

Laboratory Assessment

15

Name ______________________________

Date ______________________________

Section ______________________________

The [A] corresponds to the indicated outcome(s) found at the beginning of the laboratory exercise.

Spinal Cord, Spinal Nerves, and Meninges

Part A Assessments

Identify the features indicated in the spinal cord cross section of figure 15.8.

FIGURE 15.8 Micrograph of a spinal cord cross section with spinal nerve roots (7.5×). Label the features by placing the correct numbers in the spaces provided. [1] [2]

Posterior/Dorsal

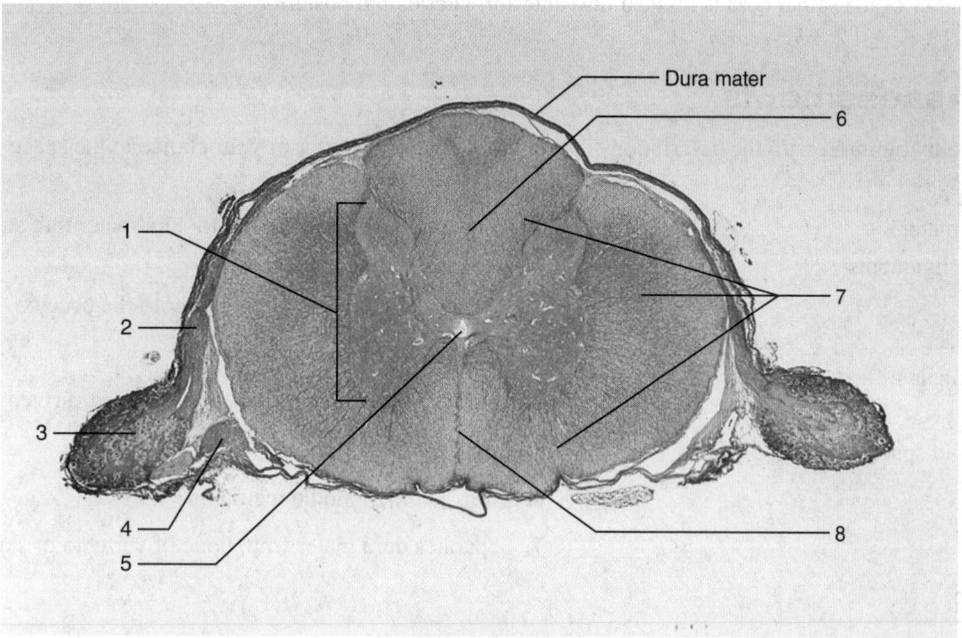

Anterior/Ventral

_____ Anterior median fissure
_____ Central canal
_____ Dorsal root ganglion
_____ Dorsal root of spinal nerve
_____ Gray matter
_____ Posterior median sulcus
_____ Ventral root of spinal nerve
_____ White matter

Part B Assessments

Complete the following statements:

1. The spinal cord gives rise to 31 pairs of _____. 🅰

2. The bulge in the spinal cord that gives off nerves to the upper limbs is called the _____ enlargement. 🅰

3. The bulge in the spinal cord that gives off nerves to the lower limbs is called the _____ enlargement. 🅰

4. The _____ is a groove that extends the length of the spinal cord posteriorly. 🅰

5. In a spinal cord cross section, the posterior _____ of the gray matter resemble the upper wings of a butterfly. 🅰

6. The motor neurons are found in the _____ roots of spinal nerves. 🅰

7. The _____ connects the gray matter on the left and right sides of the spinal cord. 🅰

8. The _____ in the gray commissure of the spinal cord contains cerebrospinal fluid and is continuous with the cavities of the brain. 🅰

9. The white matter of the spinal cord is divided into anterior, lateral, and posterior _____. 🅰

10. There are _____ pairs of cervical spinal nerves. 🅰

11. There are _____ pairs of sacral spinal nerves. 🅰

12. Cervical spinal nerve pair C1 originates between the occipital bone and the _____. 🅰

13. Spinal nerves L4 through S4 form a _____ plexus. 🅰

14. The gray matter of the spinal cord is divided into anterior, lateral, and posterior _____. 🅰

Part C Assessments

Match the terms in column A with the descriptions in column B. Place the letter of your choice in the space provided. 🅰

Column A	Column B
a. Arachnoid mater	_____ 1. Connections from pia mater to dura mater that anchor the spinal cord
b. Denticulate ligaments	
c. Dura mater	_____ 2. Inferior continuation of pia mater to the coccyx
d. Epidural space	_____ 3. Outermost layer of meninges
e. Filum terminale	_____ 4. Follows irregular contours of spinal cord surface
f. Pia mater	_____ 5. Contains cerebrospinal fluid
g. Subarachnoid space	_____ 6. Thin, weblike middle membrane
	_____ 7. Separates dura mater from bone of vertebra or skull

Reflex Arc and Reflexes

Purpose of the Exercise

To review the characteristics of reflex arcs and reflex behavior and to demonstrate some of the reflexes that occur in the human body.

Materials Needed

Rubber percussion hammer

Learning Outcomes

After completing this exercise, you should be able to

1. Demonstrate and record stretch reflexes that occur in humans.

2. Describe the components of a reflex arc.

3. Analyze the components and patterns of stretch reflexes.

Pre-Lab

Carefully read the introductory material and examine the entire lab. Be familiar with reflexes from lecture or the textbook. Answer the pre-lab questions.

Pre-Lab Questions: Select the correct answer for each of the following questions:

1. The impulse over a motor neuron will lead to
   a. an interneuron.   b. the spinal cord.
   c. a receptor.   d. an effector.
2. Stretch reflex receptors are called
   a. effectors.   b. muscle spindles.
   c. interneurons.   d. motor neurons.
3. Stretch reflexes include all of the following *except* the _____ reflex.
   a. withdrawal   b. patellar
   c. calcaneal   d. biceps
4. A withdrawal reflex could occur from
   a. striking the patellar ligament.
   b. striking the calcaneal tendon.
   c. striking the triceps tendon.
   d. touching a hot object.
5. The calcaneal reflex response is
   a. pain interpretation.
   b. flexion of the leg at the knee joint.
   c. plantar flexion of the foot.
   d. a separation of toes.
6. The quadriceps femoris is the effector muscle of the patellar reflex.
   True _____   False _____
7. The dorsal roots of spinal nerves contain the axons of the motor neurons.
   True _____   False _____
8. The normal patellar reflex response involves extension of the leg at the knee joint.
   True _____   False _____

A reflex arc represents the simplest type of nerve pathway found in the nervous system. This pathway begins with a receptor at the dendrite end of a sensory (afferent) neuron. The sensory neuron leads into the central nervous system and may communicate with one or more interneurons. Some of these interneurons, in turn, communicate with motor (efferent) neurons, whose axons (nerve fibers) lead outward to effectors. Thus, when a sensory receptor is stimulated by a change occurring inside or outside the body, impulses may pass through a reflex arc, and, as a result, effectors may respond. Such an automatic, subconscious response is called a *reflex*.

A *stretch reflex* involves a single synapse (monosynaptic) between a sensory and a motor neuron within the gray matter of the spinal cord. Examples of stretch reflexes include the patellar, calcaneal, biceps, triceps, and plantar reflexes. Other more complex *withdrawal reflexes* involve interneurons (association neurons) in combination with sensory and motor neurons; thus they are polysynaptic. Examples of withdrawal reflexes include responses to touching hot objects or stepping on sharp objects.

Reflexes demonstrated in this lab are stretch reflexes. When a muscle is stretched by a tap over its tendon, stretch receptors (proprioceptors) called *muscle spindles* are stretched within the muscle, which initiates an impulse over a reflex arc. A sensory neuron conducts an impulse from the muscle spindle into the gray matter of the spinal cord, where it synapses with a motor neuron, which conducts the impulse to the effector muscle. The stretched muscle responds by contracting to resist or reverse further stretching. These stretch reflexes are important to maintaining proper posture, balance, and movements. Observations of many of these reflexes in clinical tests on patients may indicate damage to a level of the spinal cord or peripheral nerves of the particular reflex arc.

Procedure—Reflex Arc and Reflexes

1. Study figure 16.1 as an example of a withdrawal reflex. Compare the withdrawal reflex with the stretch reflex shown in figure 16.2.
2. Work with a laboratory partner to demonstrate each of the reflexes listed. (See fig. 16.3a–e also.) *It is important that muscles involved in the reflexes be totally relaxed to observe proper responses.* If a person is trying too hard to experience the reflex or is trying to suppress the reflex, assign a multitasking activity while the stimulus with the rubber percussion hammer occurs. For example, assign a physical task with upper limbs along with a

---

FIGURE 16.1 Diagram of a withdrawal (polysynaptic) reflex arc.

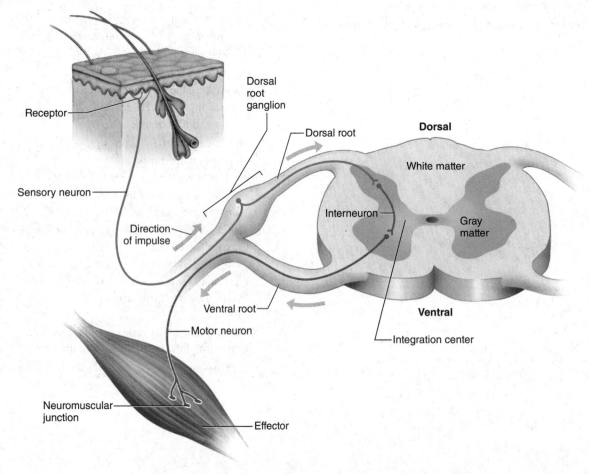

FIGURE 16.2 Diagram of a stretch (monosynaptic) reflex arc. The patellar reflex represents a specific example of a stretch reflex.

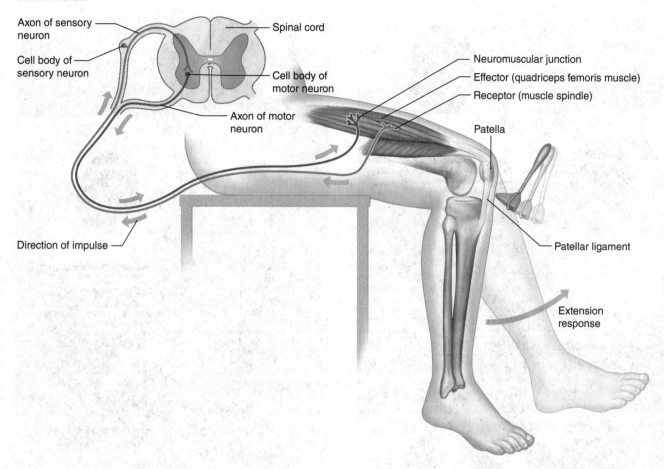

complex mental activity during the patellar reflex. After each demonstration, record your observations in the table provided in Part A of Laboratory Assessment 16.

a. *Patellar (knee-jerk) reflex.* Have your laboratory partner sit on a table (or sturdy chair) with legs relaxed and hanging freely over the edge without touching the floor. Gently strike your partner's patellar ligament (just below the patella) with the blunt side of a rubber percussion hammer (fig. 16.3*a*). The reflex arc involves the femoral nerve and the spinal cord. The normal response is a moderate extension of the leg at the knee joint.

b. *Calcaneal (ankle-jerk) reflex.* Have your partner kneel on a chair with back toward you and with feet slightly dorsiflexed over the edge and relaxed. Gently strike the calcaneal tendon (just above its insertion on the calcaneus) with the blunt side of the rubber hammer (fig. 16.3*b*). The reflex arc involves the tibial nerve and the spinal cord. The normal response is plantar flexion of the foot.

c. *Biceps (biceps-jerk) reflex.* Have your partner place a bare arm bent about 90° at the elbow on the table. Press your thumb on the inside of the elbow over the tendon of the biceps brachii, and gently strike your thumb with the rubber hammer (fig. 16.3*c*). The reflex arc involves the musculocutaneous nerve and the spi-

nal cord. Watch the biceps brachii for a response. The response might be a slight twitch of the muscle or flexion of the forearm at the elbow joint.

d. *Triceps (triceps-jerk) reflex.* Have your partner lie supine with an upper limb bent about 90° across the abdomen. Gently strike the tendon of the triceps brachii near its insertion just proximal to the olecranon process at the tip of the elbow (fig. 16.3*d*). The reflex arc involves the radial nerve and the spinal cord. Watch the triceps brachii for a response. The response might be a slight twitch of the muscle or extension of the forearm at the elbow joint.

e. *Plantar reflex.* Have your partner remove a shoe and sock and lie supine with the lateral surface of the foot resting on the table. Draw the metal tip of the rubber hammer, applying firm pressure, over the sole from the heel to the base of the large toe (fig. 16.3*e*). The normal response is flexion (curling) of the toes and plantar flexion of the foot. If the toes spread apart and dorsiflexion of the great toe occurs, the reflex is the abnormal *Babinski reflex* response (normal in infants until the nerve fibers have complete myelinization). If the Babinski reflex occurs later in life, it may indicate damage to the corticospinal tract of the CNS.

3. Complete Part B of the laboratory assessment.

FIGURE 16.3 Demonstrate each of the following reflexes: (a) patellar reflex; (b) calcaneal reflex; (c) biceps reflex; (d) triceps reflex; and (e) plantar reflex.

(a) Patellar reflex

(b) Calcaneal reflex

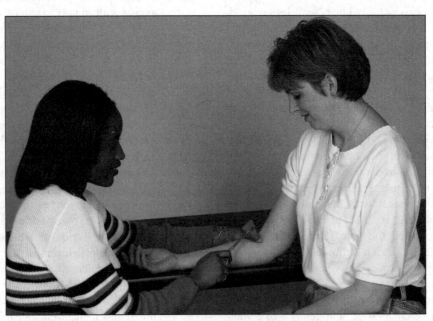

(c) Biceps reflex

FIGURE 16.3 *Continued.*

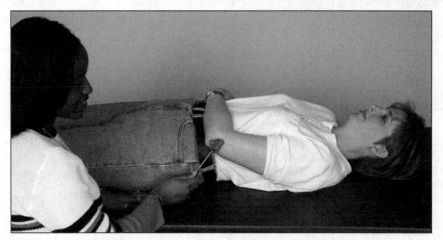

(d) Triceps reflex

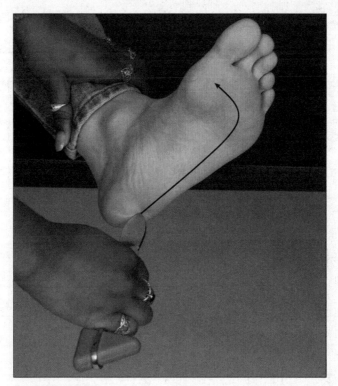

(e) Plantar reflex

Name _____

Date _____

Section _____

The 🅐 corresponds to the indicated outcome(s) found at the beginning of the laboratory exercise.

Reflex Arc and Reflexes

Part A Assessments

Complete the following table: 🅐

Reflex Tested	Response Observed	Degree of Response (hypoactive, normal, or hyperactive)	Effector Muscle Involved
Patellar			
Calcaneal			
Biceps			
Triceps			
Plantar			

Part B   Assessments

Complete the following statements:

1. A withdrawal reflex employs _____ neurons in conjunction with sensory and motor neurons. ⧸2⧹

2. Interneurons in a withdrawal reflex are located in the _____ . ⧸2⧹

3. A reflex arc begins with the stimulation of a _____ at the dendrite end of a sensory neuron. ⧸2⧹

4. Effectors of a reflex arc are glands and _____ . ⧸2⧹

5. A patellar reflex employs only _____ and motor neurons. ⧸2⧹

6. The effector muscle of the patellar reflex is the _____ . ⧸2⧹

7. The sensory stretch receptors (muscle spindles) of the patellar reflex are located in the _____ muscle. ⧸2⧹

8. The dorsal root of a spinal nerve contains the _____ neurons. ⧸2⧹

9. The normal plantar reflex results in _____ of toes. ⧸2⧹

10. Stroking the sole of the foot in infants results in dorsiflexion and toes that spread apart, called the _____ reflex. ⧸2⧹

11. List the major events that occur in the patellar reflex, from the striking of the patellar ligament to the resulting response. ⧸2⧹⧸3⧹

_____

_____

_____

Critical Thinking Assessment

What characteristics do the reflexes you demonstrated have in common?

_____

_____

_____

Brain and Cranial Nerves

Purpose of the Exercise

To review the structural and functional characteristics of the human brain and cranial nerves.

Materials Needed

Dissectible model of the human brain
Preserved human brain
Anatomical charts of the human brain

Learning Outcomes

After completing this exercise, you should be able to

1. Identify the major external and internal structures in the human brain.
2. Locate the major functional regions of the brain.
3. Identify each of the 12 pairs of cranial nerves.
4. Differentiate the functions of each cranial nerve.

Pre-Lab

Carefully read the introductory material and examine the entire lab. Be familiar with the brain and cranial nerves from lecture or the textbook. Answer the pre-lab questions.

Pre-Lab Questions: Select the correct answer for each of the following questions:

1. Each hemisphere of the cerebrum regulates
   a. motor functions on the opposite side of the body.
   b. motor functions on the same side of the body.
   c. only functions within the brain.
   d. only functions within the spinal cord.

2. There are _____ pairs of cranial nerves.
   a. 2   b. 12
   c. 31   d. 43

3. Which of the following is *not* part of the brainstem?
   a. midbrain   b. pons
   c. medulla oblongata   d. thalamus

4. The _____ is the deep lobe of the cerebrum.
   a. temporal   b. insula
   c. parietal   d. occipital

5. The _____ separates the precentral and postcentral gyrus.
   a. lateral sulcus
   b. parieto-occipital sulcus
   c. central sulcus
   d. longitudinal fissure

6. Primary vesicles of embryonic brain development include the forebrain, midbrain, and the hindbrain.
   True _____ False _____

7. Ventricles of the brain contain cerebrospinal fluid.
   True _____ False _____

8. Functions of the cerebellum include reasoning, memory, and regulation of body temperature.
   True _____ False _____

The brain, the largest and most complex part of the nervous system, contains nerve centers associated with sensory functions and is responsible for sensations and perceptions. It issues motor commands to skeletal muscles and carries on higher mental activities. It also functions to coordinate muscular movements, and it contains centers and nerve pathways necessary for the regulation of internal organs.

The cerebral cortex, comprised of gray matter and billions of interneurons, represents areas for conscious awareness and decision-making processes. Sensory areas receive information from various receptors, association areas interpret sensory input, and motor areas involve planning and controlling muscle movements. All of these functional regions are influenced and integrated together in making complex decisions. Each hemisphere primarily interprets sensory and regulates motor functions on the opposite (contralateral) side of the body.

Twelve pairs of cranial nerves arise from the ventral surface of the brain and are designated by number and name. The cranial nerves are part of the PNS; most arise from the brainstem region of the brain. Although most of these nerves conduct both sensory and motor impulses, some contain only sensory fibers associated with special sense organs. Others are primarily composed of motor fibers and are involved with the activities of muscles and glands.

Procedure A—Human Brain

The early development of the brain includes three primary vesicles formed from the neural tube: the forebrain (prosencephalon), midbrain (mesencephalon), and hindbrain (rhombencephalon). Five secondary vesicles form, including the telencephalon and diencephalon from the forebrain; the mesencephalon is retained; and the metencephalon and the myelencephalon from the hindbrain. The spaces that develop in the adult CNS from the vesicles include the central canal of the spinal cord and the ventricles of the brain. Various

TABLE 17.1 Structural Development of the Brain

Primary and Secondary Brain Vesicles	Adult Spaces Produced	Adult Brain Structures
Forebrain (prosencephalon)		
Anterior portion (telencephalon)	Lateral ventricles	Cerebrum (white matter, cortex, basal nuclei)
Posterior portion (diencephalon)	Third ventricle	Thalamus Hypothalamus Pineal gland (from epithalamus)
Midbrain (mesencephalon)	Cerebral aqueduct	Midbrain
Hindbrain (rhombencephalon)		
Anterior portion (metencephalon)	Fourth ventricle	Cerebellum Pons
Posterior portion (myelencephalon)	Fourth ventricle	Medulla oblongata

enlargements from the brain vesicles develop into the adult brain structures. See table 17.1 for a summary of the brain development.

The study of the brain will include several regional categories: the ventricles, external surface features, cerebral hemispheres, diencephalon, brainstem, and cerebellum. During each regional study, examine available anatomical charts, dissectible models, and a human brain.

1. Examine figures 17.1 and 17.2 illustrating the ventricles of the brain. The four ventricles contain a clear cerebrospinal fluid (CSF) that was secreted by blood capillaries named choroid plexuses. The CSF flows from the lateral

FIGURE 17.1 Four ventricles of the brain and their connections: (a) right lateral view and (b) anterior view.

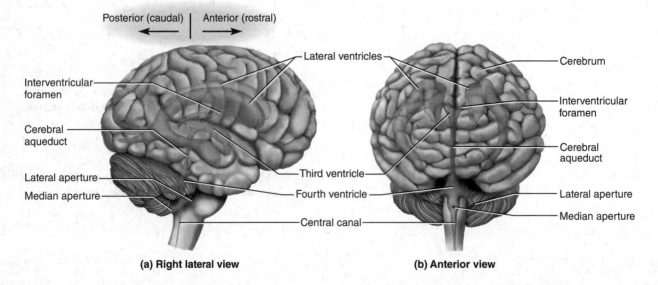

(a) Right lateral view

(b) Anterior view

FIGURE 17.2 Transverse section of the human brain showing some ventricles, surface features, and internal structures of the cerebrum (superior view). This section exposes the anterior and posterior horns of both lateral ventricles.

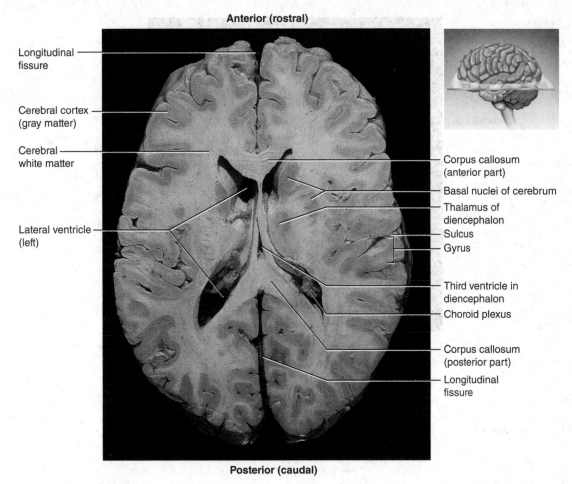

Anterior (rostral)

Longitudinal fissure

Cerebral cortex (gray matter)

Cerebral white matter

Lateral ventricle (left)

Corpus callosum (anterior part)

Basal nuclei of cerebrum

Thalamus of diencephalon

Sulcus

Gyrus

Third ventricle in diencephalon

Choroid plexus

Corpus callosum (posterior part)

Longitudinal fissure

Posterior (caudal)

ventricles into the third ventricle and the fourth ventricle and then into the central canal of the spinal cord and through pores into the subarachnoid space. Locate each of the following features using dissectible brain models:

ventricles

- lateral ventricles—the largest ventricles and one located in each cerebral hemisphere
- third ventricle—located within the diencephalon inferior to the corpus callosum
- fourth ventricle—located between the pons and the cerebellum

2. Examine figures 17.2, 17.3 and 17.4 for a study of the external surface features of the brain. Locate the following features using dissectible models and the human brain:

gyri—elevated surface ridges

- precentral gyrus
- postcentral gyrus

sulci—shallow grooves

- central sulcus—divides the frontal from the parietal lobe
- lateral sulcus (fissure)—divides the temporal from the parietal lobe

- parieto-occipital sulcus—divides the occipital from the parietal lobe

fissures—deep grooves

- longitudinal fissure—separates the cerebral hemispheres
- transverse fissure—separates the cerebrum from the cerebellum

lobes of cerebrum—names associated with bones of the cranium

- frontal lobe
- parietal lobe
- temporal lobe
- occipital lobe
- insula (insular lobe)—deep within cerebrum; not visible on surface

3. Examine figures 17.2, 17.3, 17.4, and 17.5 and table 17.2 during the study of the cerebral hemispheres. The cerebrum is the largest part of the brain and has two hemispheres connected by the corpus callosum. The lobes of the cerebrum were included in the surface features because they are also visible on the surface, except the deep insula. The functions associated with these

FIGURE 17.3 Superior view of the surface of the brain within the skull of a cadaver.

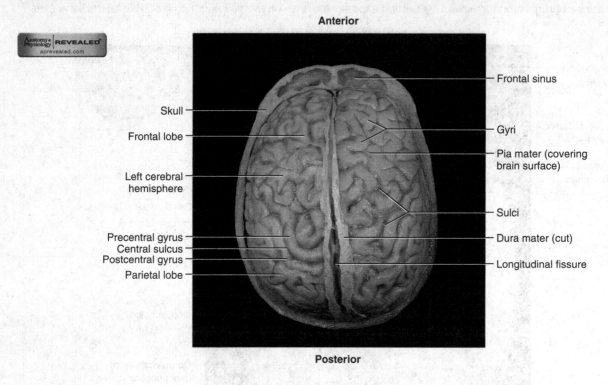

Anterior

Skull

Frontal lobe

Left cerebral hemisphere

Precentral gyrus
Central sulcus
Postcentral gyrus
Parietal lobe

Frontal sinus

Gyri

Pia mater (covering brain surface)

Sulci

Dura mater (cut)

Longitudinal fissure

Posterior

FIGURE 17.4 Lobes of the cerebrum. Retractors are used to expose the deep insula.

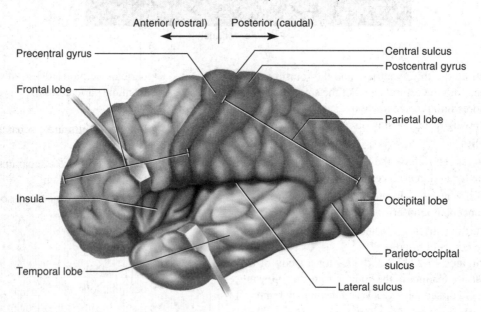

Anterior (rostral)   Posterior (caudal)

Precentral gyrus

Frontal lobe

Insula

Temporal lobe

Central sulcus
Postcentral gyrus

Parietal lobe

Occipital lobe

Parieto-occipital sulcus

Lateral sulcus

FIGURE 17.5 Diagram of a sagittal (median) section of the brain.

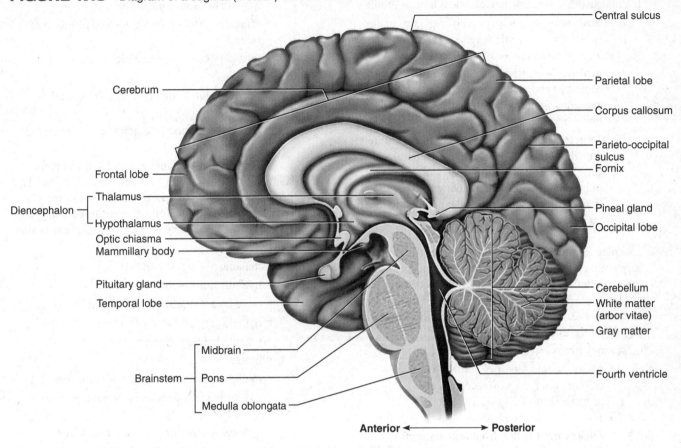

TABLE 17.2 Major Regions and Functions of the Brain

Region	Functions
1. Cerebrum	Controls higher brain functions, including sensory perception, storing memory, reasoning, and determining intelligence; initiates voluntary muscle movements
2. Diencephalon	
a. Thalamus	Relay station for sensory impulses ascending from other parts of the nervous system to the cerebral cortex
b. Hypothalamus	Helps maintain homeostasis by regulating visceral activities and by linking the nervous and endocrine systems; regulates body temperature, sleep cycles, emotions, and autonomic nervous system control
3. Brainstem	
a. Midbrain	Contains reflex centers that move the eyes and head
b. Pons	Relays impulses between higher and lower brain regions; helps regulate rate and depth of breathing
c. Medulla oblongata	Conducts ascending and descending impulses between the brain and spinal cord; contains cardiac, vasomotor, and respiratory control centers and various nonvital reflex control centers
4. Cerebellum	Processes information from other parts of the CNS by nerve tracts; integrates sensory information concerning the position of body parts; and coordinates muscle activities and maintains posture

lobes will be included in step 4. Locate the following features using dissectible models and a human brain:

cerebral cortex—thin surface layer of gray matter containing very little myelin; area of conscious awareness and processing information

cerebral white matter—largest portion of the cerebrum containing myelinated nerve fibers; transmits impulses between cerebral areas and lower brain centers

basal nuclei—masses of gray matter deep within the white matter; sometimes called basal ganglia as a clinical term; relays motor impulses from the cerebral cortex to the brainstem and spinal cord; the main structures include the caudate nucleus, putamen, and globus pallidus

4. Examine figures 17.4 and 17.6 to compare the structural lobes of the cerebrum with the functional regions of the lobes. Functional areas are not visible as distinct parts on an actual human brain. Examine the labeled areas in figure 17.6 that represent the following functional regions of the cerebrum:

sensory areas

- primary somatosensory cortex—receives information from skin receptors and proprioceptors
- somatosensory association cortex—integrates sensory information from primary cortex
- Wernicke's area—processes spoken and written language
- visual areas—consists of a primary and association (interpretation) area for vision

- auditory areas—consists of a primary and association area for hearing
- olfactory association area—interpretation of odors
- gustatory cortex—perceptions of taste

motor areas

- primary motor cortex—controls skeletal muscles
- motor association (premotor) area—planning body movements
- Broca's area—planning speech movements

5. Examine figures 17.5 and 17.7 and tables 17.1 and 17.2 during the study of the diencephalon, brainstem, and cerebellum. Locate the following features using anatomical charts, dissectible models, and a human brain:

diencephalon

- thalamus—largest portion
- hypothalamus—inferior portion
- optic chiasma (chiasm)—optic nerves meet
- mammillary bodies—pair of small humps
- pineal gland—formed from epithalamus

brainstem

- midbrain—superior region of brainstem
    - cerebral peduncles—connects pons to cerebrum
    - corpora quadrigemina—four bulges
- pons—bulge on underside of brainstem
- medulla oblongata—inferior region of brainstem

FIGURE 17.6 Some structural and functional areas of the left cerebral hemisphere. (*Note:* These areas are not visible as distinct parts of the brain.)

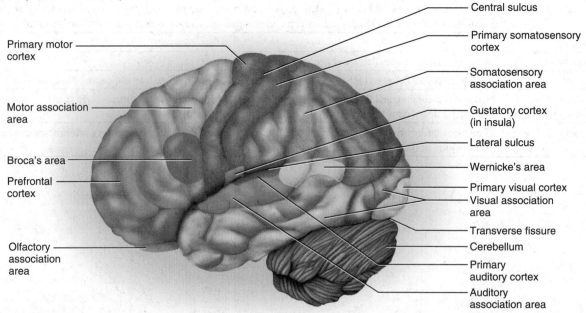

FIGURE 17.7 Cerebellum and brainstem (a) median section and (b) superior view of cerebellum.

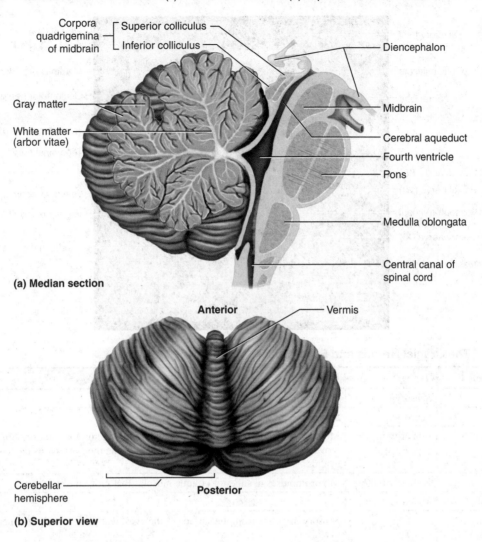

(a) Median section

(b) Superior view

cerebellum—cauliflower-like appearance
- right and left hemispheres
- vermis—connects the two hemispheres
- cerebellar cortex—gray matter portion
- arbor vitae—deeper branching pattern of white matter

6. Complete Parts A, B, and C of Laboratory Assessment 17.

Procedure B—Cranial Nerves

1. The cranial nerves are part of the PNS. Examine figure 17.8 and table 17.3.
2. Observe the model and preserved specimen of the human brain, and locate as many of the following cranial nerves as possible as you differentiate their associated functions:

olfactory nerves (I)
optic nerves (II)
oculomotor nerves (III)
trochlear nerves (IV)
trigeminal nerves (V)
abducens nerves (VI)
facial nerves (VII)
vestibulocochlear nerves (VIII)
glossopharyngeal nerves (IX)
vagus nerves (X)
accessory nerves (XI)
hypoglossal nerves (XII)

The following mnemonic device will help you learn the twelve pairs of cranial nerves in the proper order:

Old Opie **oc**casionally **tr**ies **trig**onometry, **a**nd **f**eels **v**ery **glo**omy, **vagu**e, **a**nd **hypo**active.[1]

3. Complete Parts D and E of the laboratory assessment.

[1]From *HAPS-Educator*, Winter 2002. An official publication of the Human Anatomy & Physiology Society (HAPS).

FIGURE 17.8 Photograph of the cranial nerves attached to the base of the human brain.

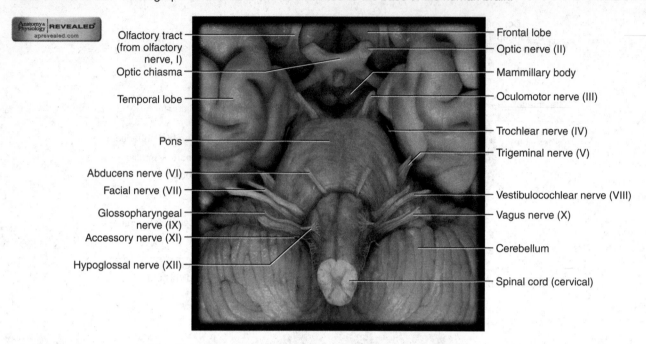

Olfactory tract (from olfactory nerve, I) · Optic chiasma · Temporal lobe · Pons · Abducens nerve (VI) · Facial nerve (VII) · Glossopharyngeal nerve (IX) · Accessory nerve (XI) · Hypoglossal nerve (XII) · Frontal lobe · Optic nerve (II) · Mammillary body · Oculomotor nerve (III) · Trochlear nerve (IV) · Trigeminal nerve (V) · Vestibulocochlear nerve (VIII) · Vagus nerve (X) · Cerebellum · Spinal cord (cervical)

TABLE 17.3   The Cranial Nerves and Functions

Number and Name		Type	Function
I	Olfactory	Sensory	Sensory impulses associated with smell
II	Optic	Sensory	Sensory impulses associated with vision
III	Oculomotor	Primarily motor*	Motor impulses to superior, inferior, and medial rectus and inferior oblique muscles that move the eyes, adjust the amount of light entering the eyes, focus the lenses, and raise the eyelids
IV	Trochlear	Primarily motor*	Motor impulses to superior oblique muscles that move the eyes
V	Trigeminal	Mixed	
	Ophthalmic division		Sensory impulses from the surface of the eyes, tear glands, scalp, forehead, and upper eyelids
	Maxillary division		Sensory impulses from the upper teeth, upper gum, upper lip, lining of the palate, and skin of the face
	Mandibular division		Sensory impulses from the scalp, skin of the jaw, lower teeth, lower gum, and lower lip. Motor impulses to muscles of mastication and to muscles in the floor of the mouth
VI	Abducens	Primarily motor*	Motor impulses to lateral rectus muscles that move the eyes laterally
VII	Facial	Mixed	Sensory impulses associated with taste receptors of the anterior tongue. Motor impulses to muscles of facial expression, tear glands, and salivary glands
VIII	Vestibulocochlear	Sensory	
	Vestibular branch		Sensory impulses associated with equilibrium
	Cochlear branch		Sensory impulses associated with hearing
IX	Glossopharyngeal	Mixed	Sensory impulses from the pharynx, tonsils, posterior tongue, and carotid arteries. Motor impulses to salivary glands and to muscles of the pharynx used in swallowing
X	Vagus	Mixed	Somatic motor impulses to muscles associated with speech and swallowing; autonomic motor impulses to the viscera of the thorax and abdomen. Sensory impulses from the pharynx, larynx, esophagus, and viscera of the thorax and abdomen
XI	Accessory	Primarily motor*	
	Cranial branch		Motor impulses to muscles of the soft palate, pharynx, and larynx
	Spinal branch		Motor impulses to muscles of the neck and shoulder
XII	Hypoglossal	Primarily motor*	Motor impulses to muscles that move the tongue

*These nerves contain a small number of sensory impulses from proprioceptors.

164

Laboratory Assessment

17

Name _____

Date _____

Section _____

The ⚠ corresponds to the indicated outcome(s) found at the beginning of the laboratory exercise.

Brain and Cranial Nerves

Part A Assessments

Match the terms in column A with the descriptions in column B. Place the letter of your choice in the space provided. ⚠

Column A	Column B
a. Central sulcus	_____ **1.** Structure formed by the crossing-over of the optic nerves
b. Cerebral cortex	_____ **2.** Part of diencephalon that forms lower walls and floor of third ventricle
c. Corpus callosum	_____ **3.** Cone-shaped gland in the upper posterior portion of diencephalon
d. Gyrus	
e. Hypothalamus	_____ **4.** Connects cerebral hemispheres
f. Insula	_____ **5.** Ridge on surface of cerebrum
g. Medulla oblongata	_____ **6.** Separates frontal and parietal lobes
h. Midbrain	_____ **7.** Part of brainstem between diencephalon and pons
i. Optic chiasma	_____ **8.** Rounded bulge on underside of brainstem
j. Pineal gland	_____ **9.** Part of brainstem continuous with the spinal cord
k. Pons	_____ **10.** Internal brain chamber filled with CSF
l. Ventricle	_____ **11.** Cerebral lobe located deep within lateral sulcus
	_____ **12.** Thin layer of gray matter on surface of cerebrum

Part B Assessments

Complete the following statements:

1. The cerebral cortex contains the _____ matter. ⚠

2. Grooves on the surface of the brain are sulci; ridges on the surface are _____. ⚠

3. The auditory areas of the brain are part of the _____ lobe. ⚠2

4. The vision areas of the brain are part of the _____ lobe. ⚠2

5. The left cerebral hemisphere primarily controls the _____ side of the body. ⚠2

6. The brainstem includes the pons, the midbrain, and the _____. ⚠

7. The delicate _____ membrane is located on the surface of the brain. ⚠

8. The _____ fissure separates the two cerebral hemispheres. ⚠

9. The primary motor cortex is located within the _____ gyrus. ⚠2

10. Arbor vitae and vermis are components of the _____. ⚠

11. The _____ ventricle is located between the pons and the cerebellum. ⚠

12. The _____ connects the two hemispheres of the cerebellum. ⚠

Part C Assessments

Identify the features indicated in the median section of the right half of the human brain in figure 17.9.

FIGURE 17.9 Label the features on this median section of the right half of the human brain by placing the correct numbers in the spaces provided.

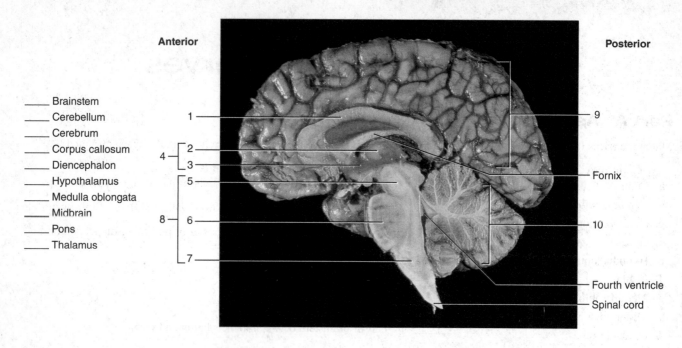

_____ Brainstem
_____ Cerebellum
_____ Cerebrum
_____ Corpus callosum
_____ Diencephalon
_____ Hypothalamus
_____ Medulla oblongata
_____ Midbrain
_____ Pons
_____ Thalamus

Part D Assessments

Identify the cranial nerves that arise from the base of the brain in figure 17.10.

FIGURE 17.10 Complete the labeling of the 12 pairs of cranial nerves as viewed from the base of the brain. The Roman numerals indicated are also often used to reference a cranial nerve. ⒊

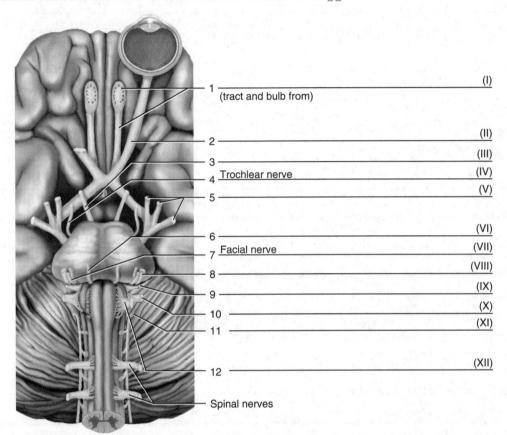

Part E Assessments

Match the cranial nerves in column A with the associated functions in column B. Place the letter of your choice in the space provided. ⒋

Column A	Column B
a. Abducens	_____ **1.** Regulates thoracic and abdominal viscera
b. Accessory	_____ **2.** Equilibrium and hearing
c. Facial	_____ **3.** Stimulates superior oblique muscle of eye
d. Glossopharyngeal	
e. Hypoglossal	_____ **4.** Sensory impulses from teeth and face
f. Oculomotor	_____ **5.** Adjusts light entering eyes and eyelid opening
g. Olfactory	_____ **6.** Smell
h. Optic	
i. Trigeminal	_____ **7.** Controls neck and shoulder movements
j. Trochlear	_____ **8.** Controls tongue movements
k. Vagus	_____ **9.** Vision
l. Vestibulocochlear	_____ **10.** Stimulates lateral rectus muscle of eye
	_____ **11.** Sensory from anterior tongue and controls salivation and secretion of tears
	_____ **12.** Sensory from posterior tongue and controls salivation and swallowing

General Senses

Purpose of the Exercise

To review the characteristics of sensory receptors and general senses and to investigate some of the general senses associated with the skin.

Materials Needed

Marking pen (washable)
Millimeter ruler
Bristle or sharp pencil
Forceps with fine points or another two-point
    discriminator device
Blunt metal probes
Three beakers (250 mL)
Warm tap water or 45°C (113°F) water bath
Cold water (ice water)
Thermometer

For Demonstration Activity:
Prepared microscope slides of tactile (Meissner's) and
    lamellated (Pacinian) corpuscles
Compound light microscope

Learning Outcomes

After completing this exercise, you should be able to

1. Associate types of sensory receptors with general senses throughout the body.
2. Determine and record the distribution of touch, warm, and cold receptors in various regions of the skin.
3. Measure the two-point threshold of various regions of the skin.

Pre-Lab

Carefully read the introductory material and examine the entire lab. Be familiar with basic structures and functions of the receptors associated with general senses from the lecture or the textbook. Answer the pre-lab questions.

Pre-Lab Questions: Select the correct answer for each of the following questions:

1. General senses include all of the following *except*
   - **a.** touch.
   - **b.** vision.
   - **c.** temperature.
   - **d.** pain.

2. Thermoreceptors are associated with
   - **a.** deep pressure.
   - **b.** light touch.
   - **c.** tissue trauma.
   - **d.** temperature changes.

3. When receptors are continuously stimulated, the sensations may fade away; this phenomenon is known as
   - **a.** tolerance.
   - **b.** perception.
   - **c.** sensory adaptation.
   - **d.** fatigue.

4. Encapsulated nerve endings include
   - **a.** tactile corpuscles.
   - **b.** pain receptors.
   - **c.** cold receptors.
   - **d.** warm receptors.

5. A lamellated corpuscle is stimulated by
   - **a.** light touch.
   - **b.** deep pressure.
   - **c.** warm temperatures.
   - **d.** cold temperatures.

6. Free nerve endings function as pain, warm, and cold receptors.
   True _____   False _____

7. Lamellated corpuscles are located in the epidermis of the skin.
   True _____   False _____

Sensory receptors are sensitive to changes that occur within the body and its surroundings. Each type of receptor is particularly sensitive to a distinct kind of environmental change and is much less sensitive to other forms of stimulation. When receptors are stimulated, they initiate nerve impulses that travel into the central nervous system. The raw form in which these receptors send information to the brain is called *sensation*. The way our brains interpret this information is called *perception*.

The sensory receptors found widely distributed throughout skin, muscles, joints, and visceral organs are associated with **general senses.** These senses include touch, pressure, temperature, pain, and the senses of muscle movement and body position. Receptors (muscle spindles) associated with muscle movements were included as part of Laboratory Exercise 16.

The general senses associated with the body surface can be classified according to the stimulus type. *Mechanoreceptors* include tactile (Meissner's) corpuscles, which are stimulated by light touch, stretch, or vibration, and lamellated (Pacinian) corpuscles, which are stimulated by deep pressure, stretch, or vibration. *Thermoreceptors* include those associated with temperature changes. Warm receptors are most sensitive to temperatures between 25°C (77°F) and 45°C (113°F). Cold receptors are most sensitive to temperatures between 10°C (50°F) and 20°C (68°F). *Nociceptors* are pain receptors that respond to tissue trauma, which may include a cut or pinch, extreme heat or cold, or excessive pressure. Tests for pain receptors are not included as part of this laboratory exercise.

Receptors possess structural differences at the dendrite ends of sensory neurons. Those with unencapsulated free nerve endings include pain, cold, and warm receptors; those with encapsulated nerve endings include tactile and lamellated corpuscles.

A sensation may fade away when receptors are continuously stimulated. This is known as *sensory adaptation,* like adjusting to a room temperature or an odor of our environment. However, sensory adaptation to pain is not as prevalent because impulses may continue into the CNS for longer periods of time. The importance of perceiving pain not only alerts us to a change in our environment and potential injury,

but it also allows us to detect body abnormalities so that proper adjustments can be taken. Unfortunately, with some spinal cord injuries or diseases such as diabetes mellitus, where the sense of pain is lost or becomes less noticeable, unwarranted tissue damages can occur.

Sensory receptors that are more specialized and confined to the head are associated with **special senses.** Laboratory Exercises 19 and 20 describe the special senses.

Procedure A—Receptors and General Senses

1. Reexamine the introduction to this laboratory exercise and study table 18.1.
2. Complete Part A of Laboratory Assessment 18.

Demonstration Activity

Observe the tactile (Meissner's) corpuscle with the microscope set up by the laboratory instructor. This type of receptor is abundant in the superficial dermis in outer regions of the body, such as in the fingertips, soles, lips, and external genital organs. It is responsible for the sensation of light touch. (See fig. 18.1.)

Observe the lamellated (Pacinian) corpuscle in the second demonstration microscope. This corpuscle is composed of many layers of connective tissue cells and has a nerve fiber in its central core. Lamellated corpuscles are numerous in the hands, feet, joints, and external genital organs. They are responsible for the sense of deep pressure (fig. 18.2). How are tactile and lamellated corpuscles similar?

_____

_____

How are they different?

_____

_____

TABLE 18.1   Receptors Associated with General Senses of Skin

Receptor Category	Respond to	Receptor Examples
Mechanoreceptors	Touch, stretch, vibration, pressure	Encapsulated tactile (Meissner's) corpuscle; encapsulated lamellated (Pacinian) corpuscle
Thermoreceptors	Temperature changes	Free nerve endings for warm temperatures; free nerve endings for cold temperatures
Nociceptors	Intense mechanical, chemical, or temperature stimuli; tissue trauma	Free nerve endings for pain

FIGURE 18.1 Tactile (Meissner's) corpuscles, such as this one, are responsible for the sensation of light touch (250×).

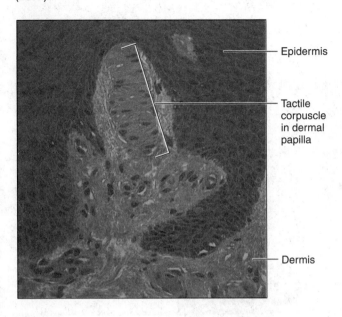

- Epidermis
- Tactile corpuscle in dermal papilla
- Dermis

FIGURE 18.2 Lamellated (Pacinian) corpuscles, such as this one, are responsible for the sensation of deep pressure (100×).

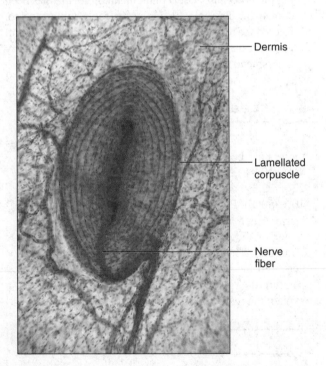

- Dermis
- Lamellated corpuscle
- Nerve fiber

Procedure B—Sense of Touch

1. Investigate the distribution of touch receptors in your laboratory partner's skin. To do this, follow these steps:
   a. Use a marking pen and a millimeter ruler to prepare a square with 2.5 cm on each side on the skin on your partner's inner wrist, near the palm.
   b. Divide the square into smaller squares with 0.5 cm on a side, producing a small grid.
   c. Ask your partner to rest the marked wrist on the tabletop and to keep his or her eyes closed throughout the remainder of the experiment.
   d. Press the end of a bristle on the skin in some part of the grid, using just enough pressure to cause the bristle to bend. A sharp pencil could be used as an alternate device.
   e. Ask your partner to report whenever the touch of the bristle is felt. Record the results in Part B of the laboratory assessment.
   f. Continue this procedure until you have tested twenty-five different locations on the grid. Move randomly through the grid to help prevent anticipation of the next stimulation site.
2. Test two other areas of exposed skin in the same manner, and record the results in Part B of the laboratory assessment.
3. Answer the questions in Part B of the laboratory assessment.

Procedure C—Two-Point Threshold

1. Test your partner's ability to recognize the difference between one and two points of skin being stimulated simultaneously. To do this, follow these steps:
   a. Have your partner place a hand with the palm up on the table and close his or her eyes.
   b. Hold the tips of a forceps tightly together and gently touch the skin on your partner's fingertip (fig. 18.3).
   c. Ask your partner to report if it feels like one or two points are touching the finger.
   d. Allow the tips of the forceps to spread so they are 1 mm apart, press both points against the skin simultaneously, and ask your partner to report as before.
   e. Repeat this procedure, allowing the tips of the forceps to spread more each time until your partner can feel both tips being pressed against the skin. The minimum distance between the tips of the forceps when both can be felt is called the *two-point threshold*. As soon as you are able to distinguish two points, two separate receptors are being stimulated instead of only one receptor (fig. 18.3).
   f. Record the two-point threshold for the skin of a fingertip in Part C of the laboratory assessment.

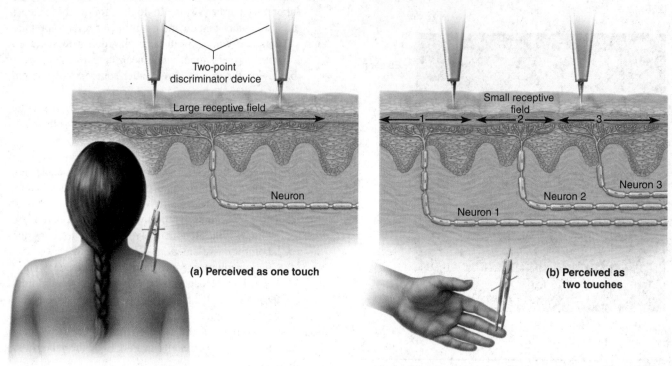

(a) Perceived as one touch

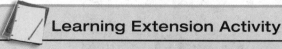

(b) Perceived as two touches

2. Repeat this procedure to determine the two-point threshold of the palm, the back of the hand, the back of the neck, the forearm, and the leg. Record the results in Part C of the laboratory assessment.

3. Answer the questions in Part C of the laboratory assessment.

Procedure D—Sense of Temperature

1. Investigate the distribution of *warm (heat) receptors* in your partner's skin. To do this, follow these steps:

   a. Mark a square with 2.5 cm sides on your partner's palm.

   b. Prepare a grid by dividing the square into smaller squares, 0.5 cm on a side.

   c. Have your partner rest the marked palm on the table and close his or her eyes.

   d. Heat a blunt metal probe by placing it in a beaker of warm (hot) water (about 40–45°C/ 104–113°F) for a minute or so. (*Be sure the probe does not get so hot that it burns the skin.*) Use a thermometer to monitor the appropriate warm water from the tap or the water bath.

   e. Wipe the probe dry and touch it to the skin on some part of the grid.

   f. Ask your partner to report if the probe feels warm. Then record the results in Part D of the laboratory assessment.

   g. Keep the probe warm, and repeat the procedure until you have randomly tested several different locations on the grid.

2. Investigate the distribution of *cold receptors* by repeating the procedure. Use a blunt metal probe that has been cooled by placing it in ice water for a minute or so. Record the results in Part D of the laboratory assessment.

3. Answer the questions in Part D of the laboratory assessment.

Learning Extension Activity

Prepare three beakers of water of different temperatures. One beaker should contain warm water (about 40°C/104°F), one should be room temperature (about 22°C/72°F), and one should contain cold water (about 10°C/50°F). Place the index finger of one hand in the warm water and, at the same time, place the index finger of the other hand in the cold water for about 2 minutes. Then, simultaneously move both index fingers into the water at room temperature. What temperature do you sense with each finger? How do you explain the resulting perceptions?

_____

_____

_____

_____

_____

_____

Name _____

Date _____

Section _____

The Ⓐ corresponds to the indicated outcome(s) found at the beginning of the laboratory exercise.

General Senses

Part A—Receptors and General Senses Assessments

Complete the following statements:

1. Whenever tissues are damaged, _____ receptors are likely to be stimulated. Ⓐ

2. Receptors that are sensitive to temperature changes are called _____. Ⓐ

3. A sensation may seem to fade away when receptors are continuously stimulated as a result of _____ adaptation. Ⓐ

4. Tactile (Meissner's) corpuscles are responsible for the sense of light _____. Ⓐ

5. Lamellated (Pacinian) corpuscles are responsible for the sense of deep _____. Ⓐ

6. _____ receptors are most sensitive to temperatures between 25°C (77°F) and 45°C (113°F). Ⓐ

7. _____ receptors are most sensitive to temperatures between 10°C (50°F) and 20°C (68°F). Ⓐ

8. Widely distributed sensory receptors throughout the body are associated with _____ senses in contrast to special senses. Ⓐ

Part B—Sense of Touch Assessments

1. Record a + to indicate where the bristle was felt and a *0* to indicate where it was not felt. Ⓐ

|←— 2.5 cm —→|

2.5 cm

Skin of wrist

2. Show the distribution of touch receptors in two other regions of skin. Ⓐ

Region tested _____   Region tested _____

3. Answer the following questions:

   a. How do you describe the pattern of distribution for touch receptors in the regions of the skin you tested? Ⓐ

   _____

   _____

   b. How does the concentration of touch receptors seem to vary from region to region? Ⓐ _____

   _____

Part C—Two-Point Threshold Assessments

1. Record the two-point threshold in millimeters for skin in each of the following regions: 3

   Fingertip _____

   Palm _____

   Back of hand _____

   Back of neck _____

   Forearm _____

   Leg _____

2. Answer the following questions:

   a. What region of the skin tested has the greatest ability to discriminate two points? 3 _____

   b. What region of the skin tested has the least sensitivity to this test? 3 _____

   c. What is the significance of these observations in questions *a* and *b*? 3 _____

   _____

   _____

Part D—Sense of Temperature Assessments

1. Record a + to indicate where warm was felt and a *0* to indicate where it was not felt. 2

   Skin of palm

2. Record a + to indicate where cold was felt and a *0* to indicate where it was not felt. 2

   Skin of palm

3. Answer the following questions:

   a. How do temperature receptors appear to be distributed in the skin of the palm? 2 _____

   _____

   b. Compare the distribution and concentration of warm and cold receptors in the skin of the palm. 2 _____

   _____

   _____

Laboratory Exercise 19

Smell and Taste

Purpose of the Exercise

To review the structures of the organs of smell and taste and to investigate the abilities of smell and taste receptors to discriminate various chemical substances.

Materials Needed

For Procedure A—Sense of Smell (Olfaction)
Set of substances in stoppered bottles: cinnamon, sage, vanilla, garlic powder, oil of clove, oil of wintergreen, and perfume

For Procedure B—Sense of Taste (Gustation)
Paper cups (small)
Cotton swabs (sterile; disposable)
5% sucrose solution (sweet)
5% NaCl solution (salt)
1% acetic acid or unsweetened lemon juice (sour)
0.5% quinine sulfate solution or 0.1% Epsom salt solution (bitter)
1% monosodium glutamate (MSG) solution (umami)

For Demonstration Activities:
Compound light microscope
Prepared microscope slides of olfactory epithelium and of taste buds

For Learning Extension Activity:
Pieces of apple, potato, carrot, and onion or packages of mixed flavors of LifeSavers

Safety

▶ Be aware of possible food allergies when selecting test solutions and foods to taste.
▶ Prepare fresh solutions for use in Procedure B.
▶ Wash your hands before starting the taste experiment.
▶ Wear disposable gloves when performing taste tests on your laboratory partner.
▶ Use a clean cotton swab for each test. Do not dip a used swab into a test solution.
▶ Dispose of used cotton swabs and paper towels as directed.
▶ Wash your hands before leaving the laboratory.

Learning Outcomes

After completing this exercise, you should be able to

1 Compare the characteristics of the smell (olfactory) receptors and the taste (gustatory) receptors.

2 Explain how the senses of smell and taste function and are subsequently preceived by the brain.

3 Record recognized odors and the time needed for olfactory sensory adaptation to occur.

4 Locate the distribution of taste receptors on the surface of the tongue and mouth cavity.

Pre-Lab

Carefully read the introductory material and examine the entire lab. Be familiar with the basic structures and functions of the receptors associated with smell and taste from the lecture or the textbook. Answer the pre-lab questions.

Pre-Lab Questions: Select the correct answer for each of the following questions:

1. Receptor cells for taste are located
   a. only on the tongue.
   b. on the tongue, oral cavity, and pharynx.
   c. in the oral cavity except the tongue.
   d. on the tongue and nasal passages.

2. Olfactory interpretation centers are located in the
   a. oral cavity.   b. inferior nose.
   c. temporal and frontal   d. brainstem
      lobes of the cerebrum

3. Which of the following is *not* considered a recognized taste?
   a. mint   b. salt
   c. sweet   d. umami

4. Taste interpretation occurs in the _____ of the cerebrum.
   a. frontal   b. temporal
   c. occipital   d. insula

5. Sour sensations are produced from
   a. acids.   b. sugars.
   c. ionized inorganic salts.   d. alkaloids.

175

The senses of smell (olfaction) and taste (gustation) are dependent upon chemoreceptors that are stimulated by various chemicals dissolved in liquids. The receptors of smell are found in the olfactory organs, which are located in the superior parts of the nasal cavity and in a portion of the nasal septum. The receptors of taste occur in the taste buds, which are sensory organs primarily found on the surface of the tongue. Chemicals are considered odorless and tasteless if receptor sites for them are absent.

Olfactory receptor cells are actually neurons, surrounded by columnar epithelial cells, with their apical ends covered with cilia, also called hair cells, embedded in the mucus of the superior nasal cavity. These neurons form the olfactory cranial nerves. In order to detect an odor, the molecules must first dissolve in the mucus before they bind to the receptor sites of the cilia (olfactory hairs). When receptor cells temporarily bind to an odorant molecule, it results in an action potential over the olfactory neurons passing through the foramina of the cribriform plate and nerve fibers in the olfactory bulb. Eventually the impulses arrive at interpreting centers located deep within the temporal lobes and the inferior frontal lobes of the cerebrum. Although humans have the ability to distinguish nearly 10,000 different odors, coded by perhaps less than 1,000 genes, various odors can result from combinations of the receptor cells stimulated. We become sensory adapted to an odor very quickly, but exposure to a different substance can be quickly noticed.

Taste receptor cells are located in taste buds on the tongue, but receptor cells are also distributed in other areas of the oral cavity and pharynx. A taste bud contains taste cells with terminal microvilli, called taste hairs, projecting through a taste pore on the epithelium of the tongue. Taste sensations are grouped into five recognized categories: sweet, sour, salt, bitter, and umami. The *sweet* sensation is produced from sugars, the *sour* sensation from acids, the *salt* sensation from ionized inorganic salts, the *bitter* sensation from alkaloids and spoiled foods, and the *umami* sensation from aspartic and glutamic acids or a derivative such as monosodium glutamate. Three cranial nerves conduct impulses from the taste buds. The facial nerve (VII) conducts sensory impulses from the anterior two-thirds of the tongue; the glossopharyngeal nerve (IX) from the pos-

terior tongue; and the vagus nerve (X) from the pharyngeal region. When a molecule binds to a receptor cell, a neurotransmitter is secreted and an action potential occurs over a sensory neuron. The action potential continues through the medulla oblongata and the thalamus, and is interpreted in the insula of the cerebrum. Sensory adaptation also occurs rather quickly, which helps explain the reason the first few bites of a particular food have the most vivid flavor.

The senses of smell and taste function closely together, because substances that are tasted often are smelled at the same moment, and they play important roles in the selection of foods. The taste and aroma of foods are also influenced by appearance, texture, temperature, and the person's mood.

Procedure A—Sense of Smell (Olfaction)

1. Reexamine the introduction to this laboratory exercise and study figure 19.1.
2. Complete Part A of Laboratory Assessment 19.
3. Test your laboratory partner's ability to recognize the odors of the bottled substances available in the laboratory. To do this, follow these steps:
   a. Have your partner keep his or her eyes closed.
   b. Remove the stopper from one of the bottles, and hold it about 4 cm under your partner's nostrils for about 2 seconds.
   c. Ask your partner to identify the odor, and then replace the stopper.
   d. Record your partner's response in Part B of the laboratory assessment.
   e. Repeat steps *b–d* for each of the bottled substances.
4. Repeat the preceding procedure, using the same set of bottled substances, but present them to your partner in a different sequence. Record the results in Part B of the laboratory assessment.
5. Wait 10 minutes and then determine the time it takes for your partner to experience olfactory sensory adaptation. To do this, follow these steps:
   a. Ask your partner to breathe in through the nostrils and exhale through the mouth.
   b. Remove the stopper from one of the bottles, and hold it about 4 cm under your partner's nostrils.
   c. Keep track of the time that passes until your partner is no longer able to detect the odor of the substance.
   d. Record the result in Part B of the laboratory assessment.
   e. Wait 5 minutes and repeat this procedure, using a different bottled substance.
   f. Test a third substance in the same manner.
   g. Record the results as before.
6. Complete Part B of the laboratory assessment.

FIGURE 19.1 Structures associated with with smell.

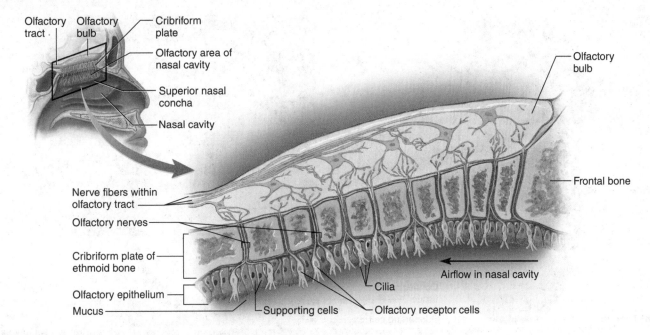

Demonstration Activity

Observe the olfactory epithelium in the demonstration microscope. The olfactory receptor cells are spindle-shaped, bipolar neurons with spherical nuclei. They also have six to eight cilia at their apical ends. The supporting cells are pseudostratified columnar epithelial cells. However, in this region the tissue lacks goblet cells. (See fig. 19.2.)

FIGURE 19.2 Olfactory receptors have cilia at their apical ends (250×).

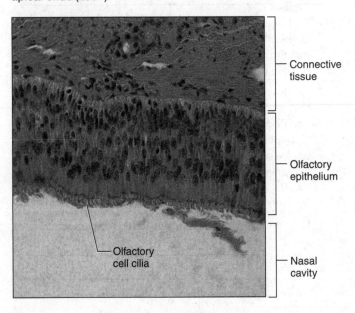

Procedure B—Sense of Taste (Gustation)

1. Reexamine the introduction to this laboratory exercise and study figure 19.3.
2. Complete Part C of the laboratory assessment.
3. Map the distribution of the receptors for the primary taste sensations on your partner's tongue. To do this, follow these steps:
   a. Ask your partner to rinse his or her mouth with water and then partially dry the surface of the tongue with a paper towel.
   b. Moisten a clean cotton swab with 5% sucrose solution, and touch several regions of your partner's tongue with the swab.
   c. Each time you touch the tongue, ask your partner to report what sensation is experienced.
   d. Test the tip, sides, and back of the tongue in this manner.
   e. Test some other representative areas inside the oral cavity such as the cheek, gums, and roof of the mouth (hard and soft palate) for a sweet sensation.
   f. Record your partner's responses in Part D of the laboratory assessment.
   g. Have your partner rinse his or her mouth and dry the tongue again, and repeat the preceding procedure, using each of the other four test solutions—NaCl, acetic acid, quinine or Epsom salt, and MSG solution. Be sure to use a fresh swab for each test substance and dispose of used swabs and paper towels as directed.
4. Complete Part D of the laboratory assessment.

FIGURE 19.3 Structures associated with taste: (a) tongue, (b) papillae, and (c) taste bud.

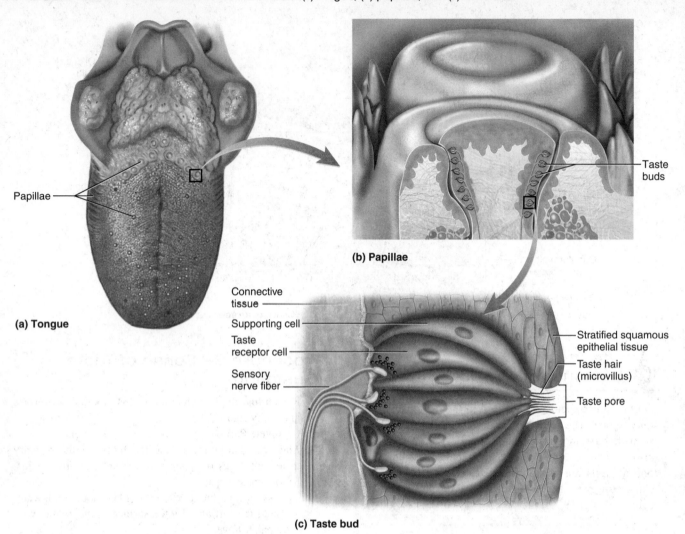

(a) **Tongue**

Papillae

(b) **Papillae**

Taste buds

Connective tissue

Supporting cell

Taste receptor cell

Sensory nerve fiber

Stratified squamous epithelial tissue

Taste hair (microvillus)

Taste pore

(c) **Taste bud**

Demonstration Activity

Observe the oval-shaped taste bud in the demonstration microscope. Note the surrounding epithelial cells. The taste pore, an opening into the taste bud, may be filled with taste hairs (microvilli). Within the taste bud there are supporting cells and thinner taste-receptor cells, which often have lightly stained nuclei (fig. 19.4).

Learning Extension Activity

Test your laboratory partner's ability to recognize the tastes of apple, potato, carrot, and onion. (A package of mixed flavors of LifeSavers is a good alternative.) To do this, follow these steps:

1. Have your partner close his or her eyes and hold the nostrils shut.
2. Place a small piece of one of the test substances on your partner's tongue.
3. Ask your partner to identify the substance without chewing or swallowing it.
4. Repeat the procedure for each of the other substances.

How do you explain the results of this experiment?

_____

_____

FIGURE 19.4 Taste receptors are found in taste buds (400×).

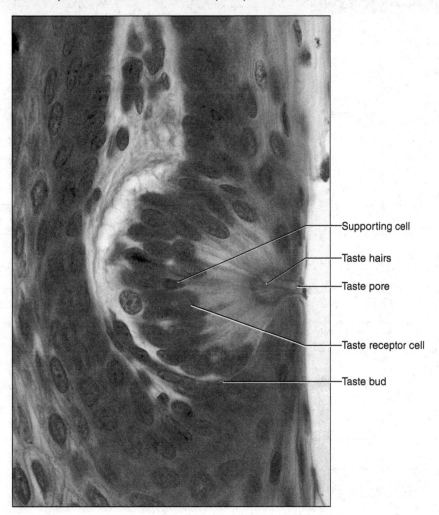

Supporting cell

Taste hairs

Taste pore

Taste receptor cell

Taste bud

NOTES

Name _____

Date _____

Section _____

The Ⓐ corresponds to the indicated outcome(s) found at the beginning of the laboratory exercise.

Smell and Taste

Part A Assessments

Complete the following statements:

1. The distal ends of the olfactory neurons are covered with hairlike _____. Ⓐ

2. Before gaseous substances can stimulate the olfactory receptors, they must be dissolved in _____ that surrounds the cilia. Ⓐ

3. The axons of olfactory receptors pass through small openings in the _____ of the ethmoid bone. Ⓐ

4. The olfactory interpreting centers are located deep within the temporal lobes and at the base of the _____ lobes of the cerebrum. Ⓐ

5. Olfactory sensations usually fade rapidly as a result of _____. Ⓐ

6. A chemical would be considered _____ if a person lacks a particular receptor site on the cilia of the olfactory neurons. Ⓐ

Part B Sense of Smell Assessments

1. Record the results (as +, if recognized; as 0, if unrecognized) from the tests of odor recognition in the following table: Ⓐ

Substance Tested	Odor Reported	
	First Trial	Second Trial

2. Record the results of the olfactory sensory adaptation time in the following table: Ⓐ

Substance Tested	Adaptation Time in Seconds

181

3. Complete the following:

   a. How do you describe your partner's ability to recognize the odors of the substances you tested? /2

   _____

   _____

   b. Compare your experimental results with those of others in the class. Did you find any evidence to indicate that individuals may vary in their abilities to recognize odors? Explain your answer. /2

   _____

   _____

   _____

Critical Thinking Assessment

Does the time it takes for sensory adaptation to occur seem to vary with the substances tested? Explain your answer. /3

_____

_____

_____

Part C   Assessments

Complete the following statements:

1. Taste is interpreted in the _____ of the cerebrum. /2

2. The opening to a taste bud is called a _____. /1

3. The _____ of a taste cell are its sensitive part. /1

4. The facial, _____, and vagus cranial nerves conduct impulses related to the sense of taste. /2

5. Substances that stimulate taste cells bind with _____ sites on the surfaces of taste hairs. /1

6. Sour receptors are mainly stimulated by _____. /1

7. Salt receptors are mainly stimulated by ionized inorganic _____. /1

8. Alkaloids usually have a _____ taste. /1

Part D Sense of Taste Assessments

1. *Taste receptor distribution.* Record a + to indicate where a taste sensation seemed to originate and a *0* if no sensation occurred when the spot was stimulated.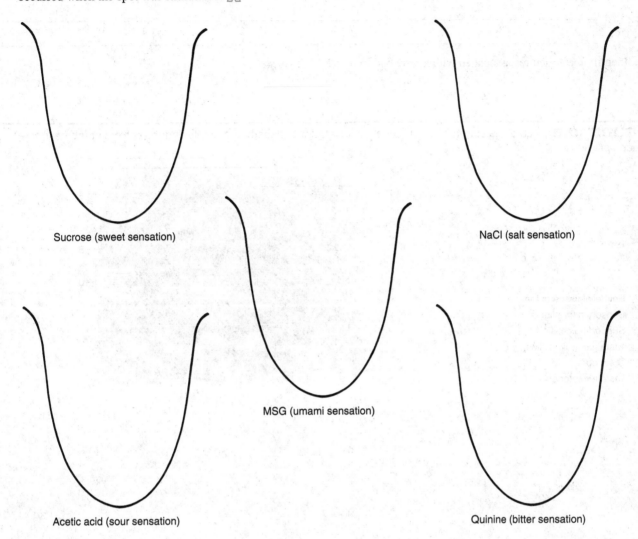

Sucrose (sweet sensation)

NaCl (salt sensation)

MSG (umami sensation)

Acetic acid (sour sensation)

Quinine (bitter sensation)

2. Complete the following:

   a. Describe how each type of taste receptor is distributed on the surface of your partner's tongue.

   _____
   _____
   _____
   _____

   b. Describe other locations inside the mouth where any sensations of sweet, salt, sour, bitter, or umami were located.

   _____
   _____
   _____
   _____

c. How does your taste distribution map on the tongue compare to those of other students in the class? 🄰

_____

_____

_____

_____

3. Identify the structures associated with a taste bud in figure 19.5. 🄰 🄰

FIGURE 19.5 Label this diagram of structures associated with a taste bud by placing the correct numbers in the spaces provided.

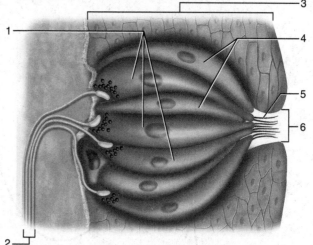

_____ Epithelial tissue of tongue
_____ Sensory nerve fibers
_____ Supporting cells
_____ Taste hair (microvillus)
_____ Taste pore
_____ Taste receptor cells

Ear and Equilibrium

Purpose of the Exercise

To review the structure and function of the organs of equilibrium and to conduct some tests of equilibrium.

Materials Needed

Swivel chair
Bright light

For Demonstration Activity:
Compound light microscope
Prepared microscope slide of semicircular duct (cross section through ampulla)

Safety

▶ Do not pick subjects who have frequent motion sickness.
▶ Have four people surround the subject in the swivel chair in case the person falls from vertigo or loss of balance.
▶ Stop your experiment if the subject becomes nauseated.

Learning Outcomes

After completing this exercise, you should be able to

1 Locate the organs of static and dynamic equilibrium and describe their functions.

2 Explain the role of vision in the maintenance of equilibrium.

3 Conduct and record the results of the Romberg and Bárány tests of equilibrium.

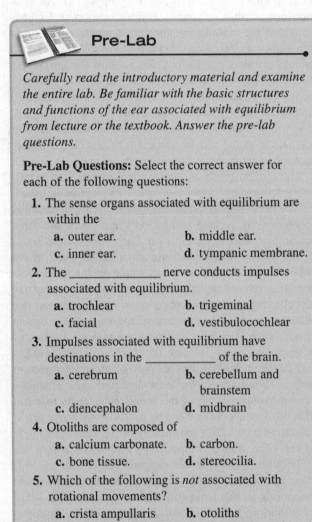

Pre-Lab

Carefully read the introductory material and examine the entire lab. Be familiar with the basic structures and functions of the ear associated with equilibrium from lecture or the textbook. Answer the pre-lab questions.

Pre-Lab Questions: Select the correct answer for each of the following questions:

1. The sense organs associated with equilibrium are within the
   a. outer ear.　　　　　b. middle ear.
   c. inner ear.　　　　　d. tympanic membrane.

2. The ＿＿＿＿＿＿ nerve conducts impulses associated with equilibrium.
   a. trochlear　　　　　b. trigeminal
   c. facial　　　　　　d. vestibulocochlear

3. Impulses associated with equilibrium have destinations in the ＿＿＿＿＿ of the brain.
   a. cerebrum　　　　　b. cerebellum and
   　　　　　　　　　　　brainstem
   c. diencephalon　　　d. midbrain

4. Otoliths are composed of
   a. calcium carbonate.　b. carbon.
   c. bone tissue.　　　　d. stereocilia.

5. Which of the following is *not* associated with rotational movements?
   a. crista ampullaris　　b. otoliths
   c. ampulla　　　　　　d. semicircular duct

6. Eye twitching movements characteristic during rotational movements are called nystagmus.
   True ＿＿＿　　False ＿＿＿

7. Otoliths are located within the semicircular ducts of the inner ear.
   True ＿＿＿　　False ＿＿＿

The sense of equilibrium involves two sets of sensory organs. One set helps to maintain the stability of the head and body when they are motionless or during linear acceleration and produces a sense of static (gravitational) equilibrium. The other set is concerned with balancing the head and body when angular acceleration produces a sense of dynamic (rotational) equilibrium.

The sense organs associated with the sense of static equilibrium are located within the vestibules of the inner ears. Two chambers, the utricle and saccule, contain receptors called maculae. Each macula is composed of hair cells embedded within a gelatinous otolithic membrane. Embedded within the otolithic membrane are numerous tiny calcium carbonate "ear stones" called otoliths. As we change our head position, gravitational forces allow the shift of the otoliths to bend the stereocilia of the hair cells of the macula, resulting in stimulation of sensory vestibular neurons. Stimulation of the maculae also occurs during linear acceleration. Horizontal acceleration (as occurs when riding in a car) involves the utricle; vertical acceleration (as occurs when riding in an elevator) involves the saccule. As a result of static equilibrium, recognition of movements such as falling are detected and postural adjustments are accomplished.

The sense organs associated with the sense of dynamic equilibrium are located within the ampullae of the three semicircular ducts of the inner ear. Each membranous semicircular duct is located within a semicircular canal of the temporal bone. A small elevation within each ampulla possesses the crista ampullaris. Each crista ampullaris contains hair cells with stereocilia embedded within a gelatinous cap called the cupula. During rotational movements, the cupula is bent, which stimulates hair cells and sensory neurons of the vestibular nerve. Because the three semicircular ducts are in different planes, rotational movements in any direction result in stimulation of the associated hair cells. Impulses from vestibular neurons of the semicircular ducts result in reflex movements of the eye. During rotational movements, characteristic twitching movements of the eyes called nystagmus occur, often accompanied by dizziness (vertigo).

Impulses from inner ear receptors travel over the vestibular neurons of the vestibulocochlear nerve and include destinations in the brainstem and cerebellum. The sense of equilibrium works subconsciously to initiate appropriate corrections to body position and movements. Additional senses work in conjunction with the inner ear and equilibrium. Stretch receptors in muscles and tendons, touch, and vision complement the inner ear for equilibrium and proper body adjustments.

Procedure A—Organs of Equilibrium

1. Re-examine the introduction to this laboratory exercise.
2. Study figures 20.1 and 20.2, which present the structures and functions of static equilibrium.
3. Study figures 20.3 and 20.4, which present the structures and functions of dynamic equilibrium.
4. Complete Part A of Laboratory Assessment 20.

Demonstration Activity

Observe the cross section of the semicircular duct through the ampulla in the demonstration microscope. Note the crista ampullaris (fig. 20.5) projecting into the lumen of the membranous labyrinth, which in a living person is filled with endolymph. The space between the membranous and bony labyrinths is filled with perilymph.

Procedure B—Tests of Equilibrium

Use figures 20.1 and 20.3 as references as you progress through the various tests of equilibrium. Perform the following tests, using a person as a test subject who is not easily disturbed by dizziness or rotational movement. Also have some other students standing close by to help prevent the test subject from falling during the tests. *The tests should be stopped immediately if the test subject begins to feel uncomfortable or nauseated.*

1. *Vision and equilibrium test.* To demonstrate the importance of vision in the maintenance of equilibrium, follow these steps:
   a. Have the test subject stand erect on one foot for 1 minute with his or her eyes open.
   b. Observe the subject's degree of unsteadiness.
   c. Repeat the procedure with the subject's eyes closed. *Be prepared to prevent the subject from falling.*
   d. Answer the questions related to the vision and equilibrium test in Part B of the laboratory assessment.
2. *Romberg test.* The purpose of this test is to evaluate how the organs of static equilibrium in the vestibule enable one to maintain balance (fig. 20.1). To conduct this test, follow these steps:
   a. Position the test subject close to a chalkboard with the back toward the board.
   b. Place a bright light in front of the subject so that a shadow of the body is cast on the board.
   c. Have the subject stand erect with feet close together and eyes staring straight ahead for 3 minutes.
   d. During the test, make marks on the chalkboard along the edge of the shadow of the subject's shoulders to indicate the range of side-to-side swaying.
   e. Measure the maximum sway in centimeters and record the results in Part B of the laboratory assessment.
   f. Repeat the procedure with the subject's eyes closed.
   g. Position the subject so one side is toward the chalkboard.
   h. Repeat the procedure with the eyes open.
   i. Repeat the procedure with the eyes closed.
      The Romberg test is used to evaluate a person's ability to integrate sensory information from proprioceptors and receptors within the organs of static equilibrium and to relay appropriate motor impulses

FIGURE 20.1 Structures of static equilibrium; (a) utricle and saccule each contain a macula; (b) macula and otolithic membrane orientation when body is upright; (c) macula and otolithic membrane changes when the body is tilted.

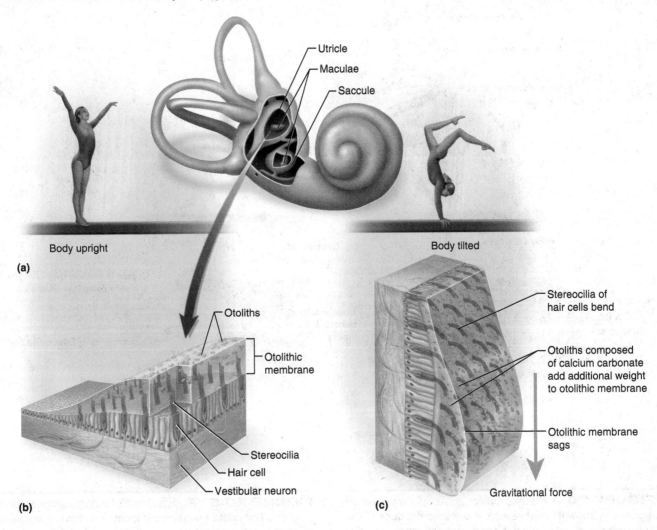

Utricle
Maculae
Saccule

Body upright
(a)

Otoliths
Otolithic membrane
Stereocilia
Hair cell
Vestibular neuron
(b)

Body tilted

Stereocilia of hair cells bend
Otoliths composed of calcium carbonate add additional weight to otolithic membrane
Otolithic membrane sags
Gravitational force
(c)

FIGURE 20.2 Particularly large otoliths are found in a freshwater drum (*Aplodinotus grunniens*). They have been used for lucky charms and jewelry. Human otoliths are microscopic size.

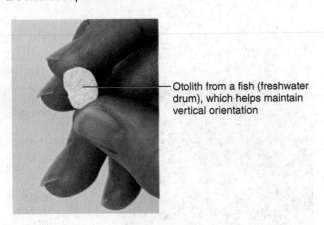

Otolith from a fish (freshwater drum), which helps maintain vertical orientation

to postural muscles. A person who shows little unsteadiness when standing with feet together and eyes open, but who becomes unsteady when the eyes are closed, has a positive Romberg test.

3. *Bárány test.* The purpose of this test is to evaluate the effects of rotational acceleration on the semicircular ducts and dynamic equilibrium (fig. 20.3). To conduct this test, follow these steps:

   a. Have the test subject sit on a swivel chair with his or her eyes open and focused on a distant object, the head tilted forward about 30°, and the hands gripped firmly to the seat. Position four people around the chair for safety. *Be prepared to prevent the subject and the chair from tipping over.*

   b. Rotate the chair ten rotations within 20 seconds.

   c. Abruptly stop the movement of the chair. The subject will still have the sensation of continuous movement and might experience some dizziness (vertigo).

FIGURE 20.3 Structures of dynamic equilibrium; (a) each semicircular duct has an ampulla containing a crista ampullaris; (b) crista ampullaris when the body is stationary; (c) changes in crista ampullaris during body rotation.

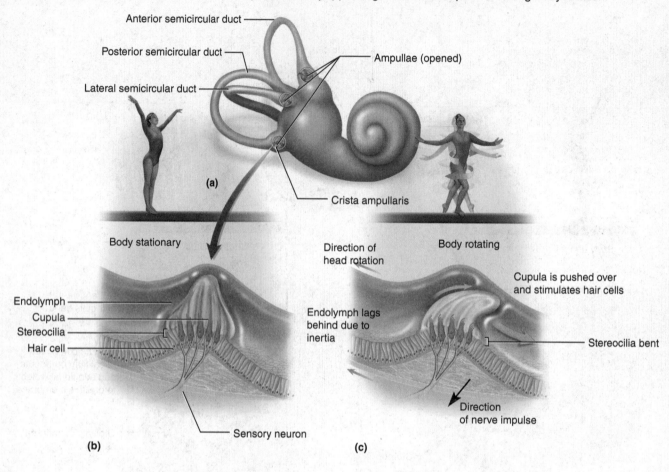

Anterior semicircular duct

Posterior semicircular duct

Lateral semicircular duct

Ampullae (opened)

Crista ampullaris

(a)

Body stationary

Endolymph
Cupula
Stereocilia
Hair cell

Sensory neuron

(b)

Direction of head rotation

Body rotating

Cupula is pushed over and stimulates hair cells

Endolymph lags behind due to inertia

Stereocilia bent

Direction of nerve impulse

(c)

FIGURE 20.4 Semicircular canal superimposed on a penny to show its relative size. A semicircular duct of the same shape would occupy the semicircular canal.

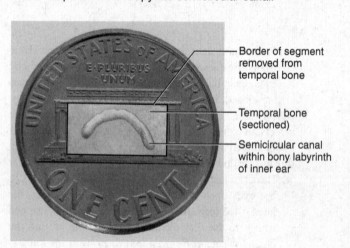

Border of segment removed from temporal bone

Temporal bone (sectioned)

Semicircular canal within bony labyrinth of inner ear

FIGURE 20.5 A micrograph of a crista ampullaris (1,400×). The crista ampullaris is located within the ampulla of each semicircular duct.

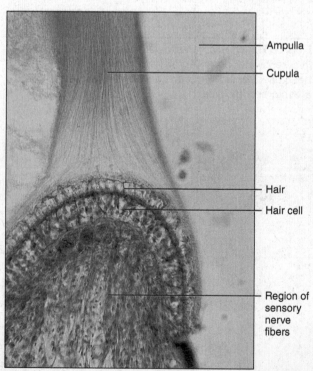

Ampulla

Cupula

Hair

Hair cell

Region of sensory nerve fibers

188

d. Have the subject look forward and immediately note the nature of the eye movements and their direction. (Such reflex eye twitching movements are called *nystagmus*.) Also note the time it takes for the nystagmus to cease. Nystagmus will continue until the cupula is returned to an original position.

e. Record your observations in Part B of the laboratory assessment.

f. Allow the subject several minutes of rest, then repeat the procedure with the subject's head tilted nearly 90° onto one shoulder.

g. After another rest period, repeat the procedure with the subject's head bent forward so that the chin is resting on the chest.

In this test, when the head is tilted about 30°, the lateral semicircular ducts receive maximal stimulation, and the nystagmus is normally from side to side. When the head is tilted at 90°, the superior ducts are stimulated, and the nystagmus is up and down. When the head is bent forward with the chin on the chest, the posterior ducts are stimulated, and the nystagmus is rotary.

4. Complete Part B of the laboratory assessment.

NOTES

Laboratory Assessment

20

Name _____

Date _____

Section _____

The Ⓐ corresponds to the indicated outcome(s) found at the beginning of the laboratory exercise.

Ear and Equilibrium

Part A Assessments

Complete the following statements:

1. The organs of static equilibrium are located within two expanded chambers within the vestibule called the _____ and the saccule. Ⓐ

2. All of the balance organs are found within the _____ bone of the skull. Ⓐ

3. Otoliths are small grains composed of _____. Ⓐ

4. Sensory impulses travel from the organs of equilibrium to the brain on vestibular neurons of the _____ nerve. Ⓐ

5. The sensory organ of a semicircular duct lies within a swelling called the _____. Ⓐ

6. The sensory organ within the ampulla of a semicircular duct is called a _____. Ⓐ

7. The _____ of this sensory organ consists of a dome-shaped gelatinous cap. Ⓐ

8. Parts of the brainstem and the _____ of the brain process impulses from the equilibrium receptors. Ⓐ

Part B Tests of Equilibrium Assessments

1. Vision and equilibrium test results:

   a. When the eyes are open, what sensory organs provide information needed to maintain equilibrium? ②

   _____

   _____

   b. When the eyes are closed, what sensory organs provide such information? ②

   _____

   _____

2. Romberg test results:

   a. Record the test results in the following table: /3\

Conditions	Maximal Movement (cm)
Back toward board, eyes open	
Back toward board, eyes closed	
Side toward board, eyes open	
Side toward board, eyes closed	

   b. Did the test subject's unsteadiness increase when the eyes were closed? _____ What is the significance of this observation? /2\ _____

   _____

   _____

   c. Why would you expect a person with impairment of the organs of equilibrium to become more unsteady when the eyes are closed? /2\ _____

   _____

   _____

3. Bárány test results:

   a. Record the test results in the following table: /3\

Position of Head	Description of Eye Movements	Time for Movement to Cease
Tilted 30° forward		
Tilted 90° onto shoulder		
Tilted forward, chin on chest		

   b. Summarize the results of this test. /3\

   _____

   _____

   _____

Critical Thinking Assessment

What additional sensory information would you expect persons with impairment of organs of equilibrium to use to supplement their relative lack of some sensory information?

_____

_____

_____

Blood Cells

Purpose of the Exercise

To review the characteristics of blood cells, to examine them microscopically, and to perform a differential white blood cell count.

Materials Needed

Compound light microscope
Prepared microscope slides of human blood (Wright's stain)
Colored pencils

For Demonstration Activity:

Mammal blood other than human or contaminant-free human blood is suggested as a substitute for collected blood
Microscope slides (precleaned)
Sterile disposable blood lancets
Alcohol swabs (wipes)
Slide staining rack and tray
Wright's stain
Distilled water

For Learning Extension Activity:

Prepared slides of pathological blood, such as eosino-philia, leukocytosis, leukopenia, and lymphocytosis

Safety

▶ It is important that students learn and practice correct procedures for handling body fluids. Consider using either mammal blood other than human or contaminant-free blood that has been tested and is available from various laboratory supply houses. Some of the procedures might be accomplished as demonstrations only. If student blood is used, it is important that students handle only their own blood.
▶ Use an appropriate disinfectant to wash the laboratory tables before and after the procedures.
▶ Wear disposable gloves and safety glasses when handling blood samples.
▶ Clean the end of a finger with an alcohol swab before the puncture is performed.
▶ The sterile blood lancet should be used only once.

▶ Dispose of used lancets and blood-contaminated items in an appropriate container (never use the wastebasket).
▶ Wash your hands before leaving the laboratory.

Learning Outcomes

After completing this exercise, you should be able to

① Identify and sketch red blood cells, five types of white blood cells, and platelets.
② Describe the structure and function of red blood cells, white blood cells, and platelets.
③ Perform and interpret the results of a differential white blood cell count.

Pre-Lab

Carefully read the introductory material and examine the entire lab. Be familiar with RBCs, WBCs, and platelets from lecture or the textbook. Answer the pre-lab questions.

Pre-Lab Questions: Select the correct answer for each of the following questions:

1. Which of the following have significant functions mainly during bleeding?
   a. red blood cells **b.** white blood cells
   c. platelets **d.** plasma
2. Which of the following is among the agranulocytes?
   a. monocyte **b.** neutrophil
   c. eosinophil **d.** basophil
3. Which white blood cell has the greatest nuclear variations?
   a. monocyte **b.** neutrophil
   c. eosinophil **d.** basophil
4. A _____ lacks a nucleus.
   a. red blood cell **b.** lymphocyte
   c. monocyte **d.** basophil
5. Which cell has a large nucleus that fills most of the cell?
   a. red blood cell **b.** platelet
   c. eosinophil **d.** lymphocyte

6. Which leukocyte is the most abundant in a normal differential count?
   a. basophil   b. monocyte
   c. neutrophil   d. lymphocyte
7. Eosinophil numbers typically increase during allergic reactions.
   True _____   False _____
8. Erythrocytes are also called granulocytes because granules are visible in their cytoplasm when using Wright's stain.
   True _____   False _____

Warning

Because of the possibility of blood infections being transmitted from one student to another if blood slides are prepared in the classroom, it is suggested that commercially prepared blood slides be used in this exercise. The instructor, however, may wish to demonstrate the procedure for preparing such a slide. Observe all safety procedures for this lab.

Demonstration Activity

To prepare a stained blood slide, follow these steps:

1. Obtain two precleaned microscope slides. Avoid touching their flat surfaces.
2. Thoroughly wash hands with soap and water and dry them with paper towels.
3. Cleanse the end of the middle finger with an alcohol swab and let the finger dry in the air.
4. Remove a sterile disposable blood lancet from its package without touching the sharp end.
5. Puncture the skin on the side near the tip of the middle finger with the lancet and properly discard the lancet.
6. Wipe away the first drop of blood with the alcohol swab. Place a drop of blood about 2 cm from the end of a clean microscope slide. Cover the lanced finger location with a bandage.
7. Use a second slide to spread the blood across the first slide, as illustrated in figure 21.1. Discard the slide used for spreading the blood in the appropriate container.
8. Place the blood slide on a slide staining rack and let it dry in the air.
9. Put enough Wright's stain on the slide to cover the smear but not overflow the slide. Count the number of drops of stain that are used.
10. After 2–3 minutes, add an equal volume of distilled water to the stain and let the slide stand for

4 minutes. From time to time, gently blow on the liquid to mix the water and stain.
11. Flood the slide with distilled water until the blood smear appears light blue.
12. Tilt the slide to pour off the water, and let the slide dry in the air.
13. Examine the blood smear with low-power magnification, and locate an area where the blood cells are well distributed. Observe these cells, using high-power magnification and then an oil immersion objective if one is available.

FIGURE 21.1 To prepare a blood smear: (a) place a drop of blood about 2 cm from the end of a clean slide; (b) hold a second slide at about a 45° angle to the first one, allowing the blood to spread along its edge; (c) push the second slide over the surface of the first so that it pulls the blood with it; (d) observe the completed blood smear. The ideal smear should be 1.5 inches in length, be evenly distributed, and contain a smooth, feathered edge.

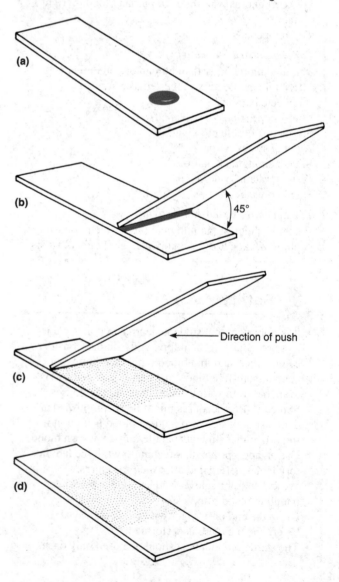

Blood is a type of connective tissue whose cells are suspended in a liquid extracellular matrix called *plasma*. Plasma is composed of water, proteins, nutrients, electrolytes, hormones, wastes, and gases. The cells, or formed elements, are mainly produced in red bone marrow, and they include *erythrocytes* (*red blood cells; RBCs*), *leukocytes* (*white blood cells; WBCs*), and some cellular fragments called *platelets* (*thrombocytes*). The formed elements compose about 45% of the total blood volume; the plasma composes approximately 55% of the blood volume.

Red blood cells contain hemoglobin and transport gases between the body cells and the lungs, white blood cells defend the body against infections, and platelets play an important role in stoppage of bleeding (hemostasis).

Clinics and hospitals test blood using a modern hematology blood analyzer. The more traditional procedures performed in these laboratory exercises will help you better understand each separate blood characteristic. On occasion, a doctor might question a blood test result from the hematology blood analyzer and request additional verification using traditional procedures.

Procedure A—Types of Blood Cells

1. Refer to figures 21.2 and 21.3 as an aid in identifying the various types of blood cells. Study the functions of the blood cells listed in table 21.1. Use the prepared slide of blood and locate each of the following:

   red blood cell (erythrocyte)
   white blood cell (leukocyte)
   - granulocytes
     - neutrophil
     - eosinophil
     - basophil

TABLE 21.1 Cellular Components of Blood

Component	Function
Red blood cell (erythrocyte)	Transports oxygen and carbon dioxide
White blood cell (leukocyte)	Destroys pathogenic microorganisms and parasites, removes worn cells, and provides immunity
Granulocytes—have granular cytoplasm	
1. Neutrophil	Phagocytizes bacteria
2. Eosinophil	Destroys parasites and helps control inflammation and allergic reactions
3. Basophil	Releases heparin (an anticoagulant) and histamine (a blood vessel dilator)
Agranulocytes—lack granular cytoplasm	
1. Monocyte	Phagocytizes dead or dying cells and microorganisms
2. Lymphocyte	Provides immunity
Platelet (thrombocyte)	Helps control blood loss from injured blood vessels; needed for blood clotting

   - agranulocytes
     - lymphocyte
     - monocyte

   platelet (thrombocyte)

2. In Part A of Laboratory Assessment 21, prepare sketches of single blood cells to illustrate each type. Pay particular attention to relative size, nuclear shape, and color of granules in the cytoplasm (if present). The sketches should be accomplished using either the high-power

FIGURE 21.2 Micrograph of a blood smear using Wright's stain (500×).

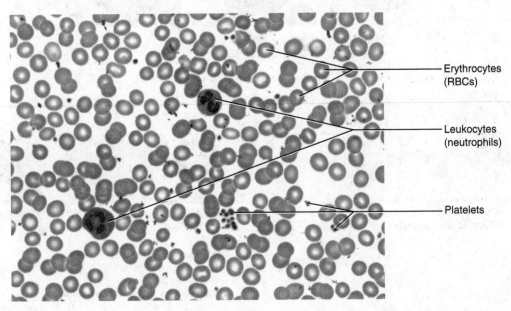

Erythrocytes (RBCs)

Leukocytes (neutrophils)

Platelets

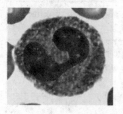 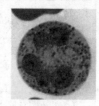  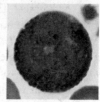 

Neutrophils (3 of many variations)
•Fine light-purple granules
•Nucleus single to five lobes (highly variable)
•Immature neutrophils, called bands, have a
 single C-shaped nucleus
•Mature neutrophils, called segs, have a lobed nucleus
•Often called polymorphonuclear leukocytes when older

Eosinophils (3 of many variations)
•Coarse reddish granules
•Nucleus usually bilobed

    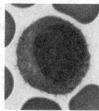

Basophils (3 of many variations)
•Coarse deep blue to almost black granules
•Nucleus often almost hidden by granules

Lymphocytes (3 of many variations)
•Slightly larger than RBCs
•Thin rim of nearly clear cytoplasm
•Nearly round nucleus appears to fill most of cell
 in smaller lymphocytes
•Larger lymphocytes hard to distinguish from monocytes

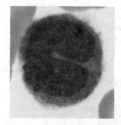 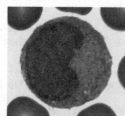 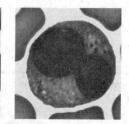

Monocytes (3 of many variations)
•Largest WBC; 2–3x larger than RBCs
•Cytoplasm nearly clear
•Nucleus round, kidney-shaped, oval, or lobed

     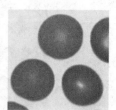

Platelets (several variations)
•Cell fragments
•Single to small clusters

Erythrocytes (several variations)
•Lack nucleus (mature cell)
•Biconcave discs
•Thin centers appear almost hollow

196

objective or the oil immersion objective of the compound light microscope.

3. Complete Part B of the laboratory assessment.

Procedure B—Differential White Blood Cell Count

A differential white blood cell count is performed to determine the percentage of each of the various types of white blood cells present in a blood sample. The test is useful because the relative proportions of white blood cells may change in particular diseases as indicated in table 21.2. Neutrophils, for example, usually increase during bacterial infections, whereas eosinophils may increase during certain parasitic infections and allergic reactions.

1. To make a differential white blood cell count, follow these steps:

   a. Using high-power magnification or an oil immersion objective, focus on the cells at one end of a prepared blood slide where the cells are well distributed.

   b. Slowly move the blood slide back and forth, following a path that avoids passing over the same cells twice (fig. 21.4).

TABLE 21.2 Differential White Blood Cell Count

Cell Type	Normal Value (percent)	Elevated Levels May Indicate
Neutrophil	54–62	Bacterial infections, stress
Lymphocyte	25–33	Mononucleosis, whooping cough, viral infections
Monocyte	3–9	Malaria, tuberculosis, fungal infections
Eosinophil	1–3	Allergic reactions, autoimmune diseases, parasitic worms
Basophil	<1	Cancers, chicken pox, hypothyroidism

c. Each time you encounter a white blood cell, identify its type and record it in Part C of the laboratory assessment.

d. Continue searching for and identifying white blood cells until you have recorded 100 cells in the data table. *Percent* means "parts of 100" for each type of white blood cell, so the total number observed is equal to its percentage in the blood sample.

2. Complete Part C of the laboratory assessment.

FIGURE 21.4 Move the blood slide back and forth to avoid passing the same cells twice.

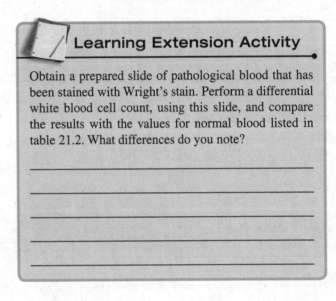

Learning Extension Activity

Obtain a prepared slide of pathological blood that has been stained with Wright's stain. Perform a differential white blood cell count, using this slide, and compare the results with the values for normal blood listed in table 21.2. What differences do you note?

_____

_____

_____

_____

_____

Name _____

Date _____

Section _____

The  corresponds to the indicated outcome(s) found at the beginning of the laboratory exercise.

Blood Cells

Part A   Assessments

Sketch a single blood cell of each type in the following spaces. Use colored pencils to represent the stained colors of the cells. Label any features that can be identified. ⚠

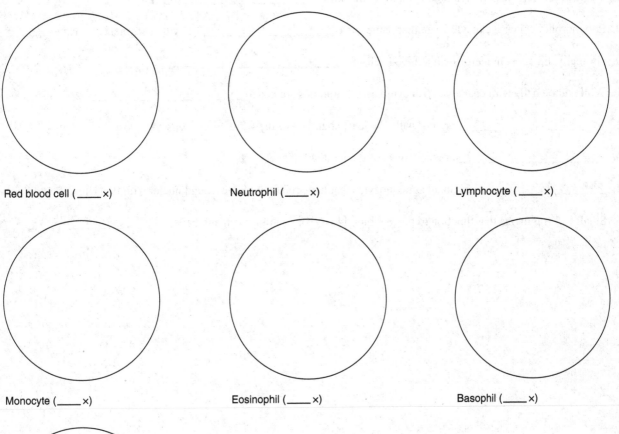

Red blood cell (____ ×)   Neutrophil (____ ×)   Lymphocyte (____ ×)

Monocyte (____ ×)   Eosinophil (____ ×)   Basophil (____ ×)

Platelet (____ ×)

Part B Assessments

Complete the following statements:

1. Mature red blood cells are also called _____. 🄰

2. The shape of a red blood cell can be described as a _____ disc. 🄰

3. The functions of red blood cells are _____. 🄰

4. _____ is the oxygen-carrying substance in a red blood cell. 🄰

5. A mature red blood cell cannot reproduce because it lacks the _____ that was extruded during late development. 🄰

6. White blood cells are also called _____. 🄰

7. White blood cells with granular cytoplasm are called _____. 🄰

8. White blood cells lacking granular cytoplasm are called _____. 🄰

9. Polymorphonuclear leukocyte is another name for a _____ with a segmented nucleus. 🄰

10. Normally, the most numerous white blood cells are _____. 🄰

11. White blood cells with coarse reddish cytoplasmic granules are called _____. 🄰

12. _____ are normally the least abundant of the white blood cells. 🄰

13. _____ are the largest of the white blood cells. 🄰

14. _____ are small agranulocytes that have relatively large, round nuclei with thin rims of cytoplasm. 🄰

15. Small cell fragments that function to prevent blood loss from an injury site are called _____. 🄰

Part C Assessments

1. *Differential White Blood Cell Count Data Table*. As you identify white blood cells, record them on the table by using a tally system, such as 𝍸 ||. Place tally marks in the "Number Observed" column and total each of the five WBCs when the differential count is completed. Obtain a total of all five WBCs counted to determine the percent of each WBC type. 3

Type of WBC	Number Observed	Total	Percent
Neutrophil			
Lymphocyte			
Monocyte			
Eosinophil			
Basophil			
		Total of column	

2. How do the results of your differential white blood cell count compare with the normal values listed in table 21.2? 3

____

____

____

____

____

Critical Thinking Assessment

What is the difference between a differential white blood cell count and a total white blood cell count?

____

____

____

____

____

3. Identify the blood cells indicated in figure 21.5.

FIGURE 21.5 Label the specific blood cells on this micrograph of a stained blood smear (400×). 🔺

1 _____

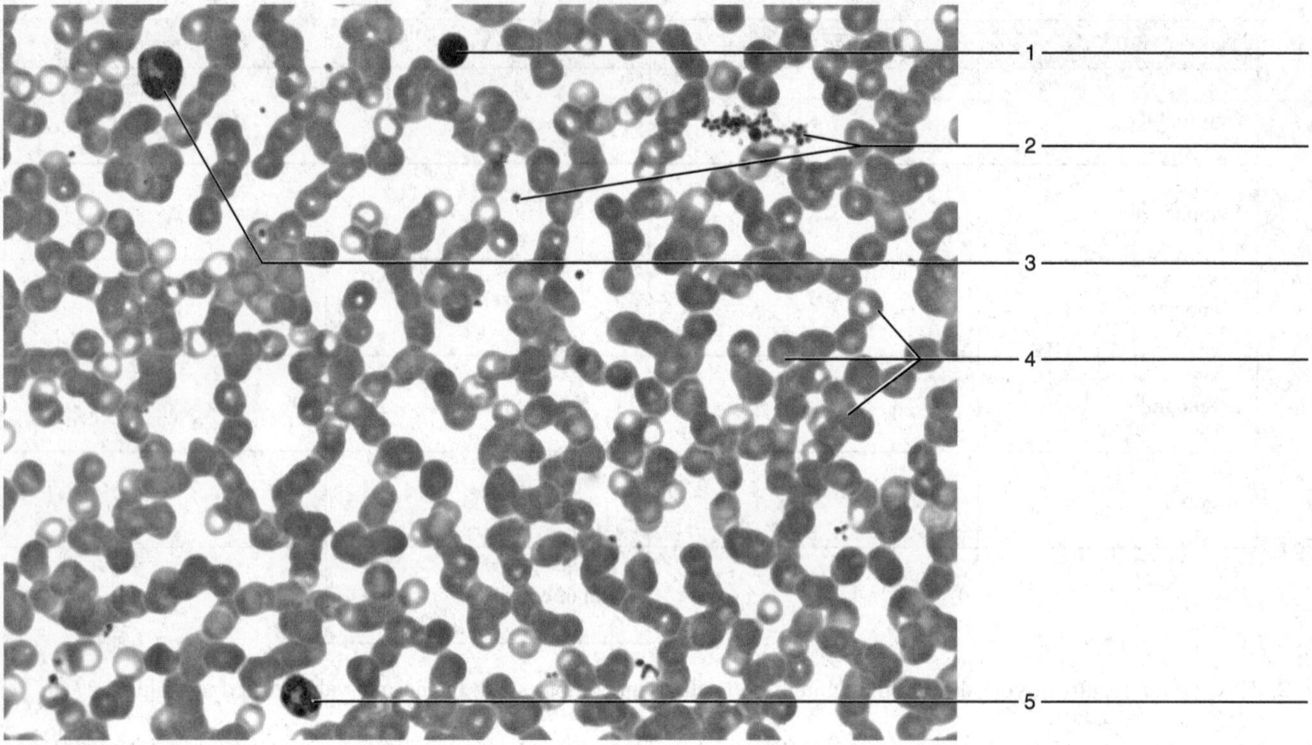

2 _____

3 _____

4 _____

5 _____

Blood Typing

Purpose of the Exercise

To determine the ABO blood type of a blood sample and to observe an Rh blood-typing test.

Materials Needed

For Procedure A:
ABO blood-typing kit
Simulated blood-typing kits are suggested as a
  substitute for collected blood

For Procedure B:
Microscope slide
Alcohol swabs (wipes)
Sterile blood lancet
Toothpicks
Anti-D serum
Slide warming box (Rh blood-typing box or Rh
  view box)

Safety

► It is important that students learn and practice correct procedures for handling body fluids. Consider using simulated blood-typing kits or contaminant-free blood that has been tested and is available from various laboratory supply houses. Some of the procedures might be accomplished as demonstrations only. If student blood is used, it is important that students handle only their own blood.
► Use an appropriate disinfectant to wash the laboratory tables before and after the procedures.
► Wear disposable gloves and safety goggles when handling blood samples.
► Clean the end of a finger with an alcohol swab before the puncture is performed.
► The sterile blood lancet should be used only once.
► Dispose of used lancets and blood-contaminated items in an appropriate container (never use the wastebasket).
► Wash your hands before leaving the laboratory.

Learning Outcomes

After completing this exercise, you should be able to

1. Analyze the basis of ABO blood typing.
2. Interpret the ABO type of a blood sample.
3. Explain the basis of Rh blood typing.
4. Interpret the Rh type of a blood sample.

Pre-Lab

Carefully read the introductory material and examine the entire lab. Be familiar with the ABO blood group and the Rh blood group from lecture or the textbook. Answer the pre-lab questions.

Pre-Lab Questions: Select the correct answer for each of the following questions:

1. The antigens related to the ABO blood group are located
   a. on the red blood cell membrane.
   b. within the red blood cell nucleus.
   c. within the red blood cell cytosol.
   d. on the red blood cell ribosome.

2. The D antigen related to the Rh factor is present in about _____ of Americans.
   a. 4 %   b. 38 %
   c. 47 %   d. 85 %

3. Blood type _____ is considered the universal donor within the ABO blood group.
   a. A   b. B
   c. AB   d. O

4. Blood type _____ is considered the universal recipient within the ABO blood group.
   a. A   b. B
   c. AB   d. O

5. Hemolytic disease of the newborn could be of concern when an
   a. Rh-positive fetus and an Rh-positive mother condition occur.
   b. Rh-positive fetus and an Rh-negative mother condition occur.
   c. Rh-negative fetus and an Rh-positive mother condition occur.
   d. Rh-negative fetus and an Rh-negative mother condition occur.
6. Blood type B is the most common blood type found in the United States population.
   True _____ False _____
7. An individual with blood type O lacks both RBC antigens A and B.
   True _____ False _____

Blood typing involves identifying protein substances called *antigens* present in red blood cell membranes. Although there are many different antigens associated with human red blood cells, only a few of them are of clinical importance. These include the antigens of the ABO group and those of the Rh group.

To determine which antigens are present, a blood sample is mixed with blood-typing sera that contain known types of antibodies. If a particular antibody contacts a corresponding antigen, a reaction occurs, and the red blood cells clump together (agglutination). Thus, if blood cells are mixed with serum containing antibodies that react with antigen A and the cells clump together, antigen A must be present in those cells.

Although there are many antigens on the red blood cell membranes other than those of the ABO group and the Rh group, these antigens are of the greatest concern during transfusions. Most other antigens cause little if any transfusion reactions. As a final check on blood compatibility before transfusions, an important cross-matching test occurs. In this test, small samples of the blood from the donor and the recipient are mixed to be sure clumping (agglutination) does not create a transfusion reaction. If surgery is elective, autologous transfusions are promoted, allowing some of the person's own predonated blood from storage to be used during the surgery.

An antigen commonly found on the human RBC membrane was first identified in rhesus monkeys, and it became known as the Rh factor. Although many different types of antigens are related to the Rh factor, only the D antigen is checked using an anti-D reagent. About 85% of Americans possess the D antigen and are therefore considered to be Rh-positive (Rh^+). Those lacking the D antigen are considered Rh-negative (Rh^-). If an Rh-negative woman is pregnant carrying an Rh-positive fetus, the mother might obtain some RBCs from the fetus during the birth process or during a miscarriage. As a result, the mother would begin producing anti-D antibodies, creating complications for the second and future pregnancies. This condition is known as hemolytic disease of the fetus and newborn (erythroblastosis fetalis). This condition can be prevented with the proper administration of RhoGAM, which prevents the mother from producing anti-D antibodies.

Warning

Because of the possibility of blood infections being transmitted from one student to another if blood testing is performed in the classroom, it is suggested that commercially prepared blood-typing kits containing virus-free human blood be used for ABO blood typing. The instructor may wish to demonstrate Rh blood typing. Observe all of the safety procedures listed for this lab.

Procedure A—ABO Blood Typing

1. Reexamine the introductory material and study table 22.1.
2. Compete Part A of Laboratory Assessment 22.
3. Perform the ABO blood type test using the blood-typing kit. To do this, follow these steps:
   a. Obtain a clean microscope slide and mark across its center with a wax pencil to divide it into right and left halves. Also write "Anti-A" near the edge of the left half and "Anti-B" near the edge of the right half (fig. 22.1).

TABLE 22.1 Antigens and Antibodies of the ABO Blood Group and Preferred, Permissible, and Incompatible Donors

Blood Type	RBC Antigens (Agglutinogens)	Plasma Antibodies (Agglutinins)	Preferred Donor Type	Permissible Donor Type in Limited Amounts	Incompatible Donor
A	A	Anti-B	A	O	B, AB
B	B	Anti-A	B	O	A, AB
AB (universal recipient)	A and B	Neither anti-A nor anti-B	AB	A, B, O	None
O (universal donor)	Neither A nor B	Both anti-A and anti-B	O	No alternative types	A, B, AB

FIGURE 22.1 Slide prepared for ABO blood typing.

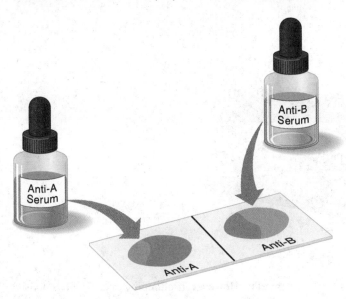

FIGURE 22.2 Four possible results of the ABO test.

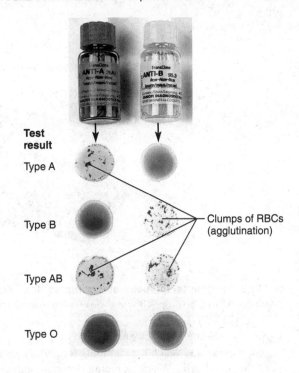

Clumps of RBCs (agglutination)

b. Place a small drop of blood on each half of the microscope slide. Work quickly so that the blood will not have time to clot.

c. Add a drop of anti-A serum to the blood on the left half and a drop of anti-B serum to the blood on the right half. Note the color coding of the anti-A and anti-B typing sera. To avoid contaminating the serum, avoid touching the blood with the serum while it is in the dropper; instead allow the serum to fall from the dropper onto the blood.

d. Use separate toothpicks to stir each sample of serum and blood together, and spread each over an area about as large as a quarter. Dispose of toothpicks in an appropriate container.

e. Examine the samples for clumping of blood cells (agglutination) after 2 minutes.

f. See table 22.2 and figure 22.2 for aid in interpreting the test results.

g. Discard contaminated materials as directed by the laboratory instructor.

4. Record your results and complete Part B of the laboratory assessment.

Critical Thinking Activity

Judging from your observations of the blood-typing results, suggest which components in the anti-A and anti-B sera caused clumping (agglutination).

_____

_____

_____

TABLE 22.2 ABO Blood-Typing Sera Reactions and U.S. Blood Type Frequency

Possible Reactions		Blood Type	U.S. Frequency (average)
Anti-A Serum	**Anti-B Serum**		
Clumping (agglutination)	No clumping	Type A	40%
No clumping	Clumping	Type B	10%
Clumping	Clumping	Type AB	4%
No clumping	No clumping	Type O	46%

Note: The inheritance of the ABO blood groups is described in Laboratory Exercise 61.

Procedure B—Rh Blood Typing

1. Reexamine the introductory material.

2. Complete Part C of the laboratory assessment.

3. To determine the Rh blood type of a blood sample, follow these steps:

a. Lance the tip of a finger. (See the demonstration procedures in Laboratory Exercise 21 for directions.) Place a small drop of blood in the center of a clean microscope slide. Cover the lanced finger location with a bandage.

b. Add a drop of anti-D serum to the blood and mix them together with a clean toothpick.

FIGURE 22.3 Slide warming box used for Rh blood typing.

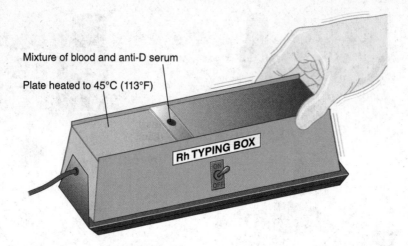

Mixture of blood and anti-D serum

Plate heated to 45°C (113°F)

Rh TYPING BOX

ON OFF

c. Place the slide on the plate of a warming box (Rh blood-typing box or Rh view box) prewarmed to 45°C (113°F) (fig. 22.3).

d. Slowly rock the box back and forth to keep the mixture moving and watch for clumping (agglutination) of the blood cells. When clumping occurs in anti-D serum, the clumps usually are smaller than those that appear in anti-A or anti-B sera, so they may be less obvious. However, if clumping occurs, the blood is called Rh positive; if no clumping occurs *within 2 minutes,* the blood is called Rh negative.

e. Discard all contaminated materials in appropriate containers.

4. Record your results and complete Part D of the laboratory assessment.

Laboratory Assessment

22

Name _____

Date _____

Section _____

The △ corresponds to the indicated outcome(s) found at the beginning of the laboratory exercise.

Blood Typing

Part A Assessments

Complete the following statements:

1. The antigens of the ABO blood group are located in the red blood cell _____. △

2. The blood of every person contains one of (how many possible?) _____ combinations of antigens. △

3. Type A blood contains antigen _____. △

4. Type B blood contains antigen _____. △

5. Type A blood contains _____ antibody in the plasma.

6. Type B blood contains _____ antibody in the plasma. △

7. Persons with ABO blood type _____ are often called universal recipients. △

8. Persons with ABO blood type _____ are often called universal donors. △

Part B Assessments

Complete the following:

1. What was the ABO type of the blood tested? △ _____

2. What ABO antigens are present in the red blood cells of this type of blood? △ _____

3. What ABO antibodies are present in the plasma of this type of blood? △ _____

4. If a person with this blood type needed a blood transfusion, what ABO type(s) of blood could be received safely? △

_____

5. If a person with this blood type was serving as a blood donor, what ABO blood type(s) could receive the blood safely? △

_____

Part C Assessments

Complete the following statements:

1. The Rh blood group was named after the _____. △

2. Of the antigens in the Rh group, the most important is _____. △

3. If red blood cells lack Rh antigens, the blood is called _____. △

4. If an Rh-negative person who is sensitive to Rh-positive blood receives a transfusion of Rh-positive blood, the donor's cells are likely to _____. △

5. An Rh-negative woman who might be carrying an _____ fetus is given an injection of RhoGAM to prevent hemolytic disease of the fetus and newborn. △

Part D Assessments

Complete the following:

1. What was the Rh type of the blood tested? 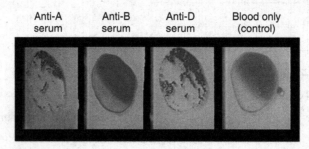 _____

2. What Rh antigen is present in the red blood cells of this type of blood? _____

3. If a person with this blood type needed a blood transfusion, what type of blood could be received safely?

_____

4. If a person with this blood type was serving as a blood donor, a person with what type of blood could receive the blood safely? _____

5. Observe the test results shown in figure 22.4 from an individual being tested for the ABO type and the Rh type.

   a. This individual has ABO type _____ .

   b. This individual has Rh type _____ .

FIGURE 22.4 Blood test results after adding anti-A serum, anti-B serum, and anti-D serum to separate drops of blood. An unaltered drop of blood is shown as a control; it does not indicate any clumping (agglutination) reaction.

Anti-A serum   Anti-B serum   Anti-D serum   Blood only (control)

Cardiac Cycle

Purpose of the Exercise

To review the events of a cardiac cycle, to become acquainted with normal heart sounds, and to record an electrocardiogram.

Materials Needed

For Procedure A—Heart Sounds:
Stethoscope
Alcohol swabs (wipes)

For Procedure B—Electrocardiogram:
Electrocardiograph (or other instrument for recording an ECG)
Cot or table
Alcohol swabs (wipes)
Electrode cream (paste)
Plate electrodes and cables
Self-sticking leads (optional)
Lead selector switch

Learning Outcomes

After completing this exercise, you should be able to

1. Interpret the major events of a cardiac cycle.
2. Associate the sounds produced during a cardiac cycle with the valves closing.
3. Correlate the components of a normal ECG pattern with the phases of a cardiac cycle.
4. Record and interpret an electrocardiogram.
5. Diagram and label the waves of an electrocardiogram (ECG) and correlate with the heart sounds.
6. Integrate the electrical changes and the heart sounds with the cardiac cycle.

Pre-Lab

Carefully read the introductory material and examine the entire lab. Be familiar with the cardiac cycle, heart sounds, cardiac conduction system, and electrocardiograms from lecture or the textbook. Answer the pre-lab questions.

Pre-Lab Questions: Select the correct answer for each of the following questions:

1. The _____ of the conduction system is known as the pacemaker.
   a. SA node   **b.** AV node
   c. AV bundle   **d.** left bundle branch

2. The _____ of the conduction system is/are located throughout the ventricular walls.
   a. AV node   **b.** left bundle branch
   c. right bundle branch   **d.** Purkinje fibers

3. The first of two heart sounds (lubb) occurs when the
   a. AV valves open.
   b. AV valves close.
   c. semilunar valves open.
   d. semilunar valves close.

4. One cardiac cycle would consist of
   a. left chamber contractions followed by right chamber contractions.
   b. right chamber contractions followed by left chamber contractions.
   c. atrial chamber contractions followed by ventricular chamber contractions.
   d. ventricular chamber contractions followed by atrial chamber contractions.

5. The depolarization of ventricular fibers is indicated by the _____ of an ECG.
   a. P wave   **b.** QRS complex
   c. T wave   **d.** P-Q interval

6. The dupp sound occurs when the semilunar valves are closing during ventricular diastole.
   True _____   False _____

7. The P wave of an ECG occurs during the repolarization of the atria.
   True _____   False _____

A set of atrial contractions while the ventricular walls relax, followed by ventricular contractions while the atrial walls relax, constitutes a *cardiac cycle*. Such a cycle is accompanied by blood pressure changes within the heart chambers, movement of blood in and out of the chambers, and opening and closing of heart valves. These valves closing produce vibrations in the tissues and thus create the sounds associated with the heartbeat. If backflow of blood occurs when a heart valve is closed, it creates a turbulence noise known as a murmur.

The regulation and coordination of the cardiac cycle involves the *cardiac conduction system*. The pathway of electrical signals originates from the *sinoatrial (SA) node* located in the right atrium near the entrance of the superior vena cava. The stimulation of the heartbeat and the heart rate originate from the SA node, so it is often called the *pacemaker*. As the signals pass through the atrial walls toward the *atrioventricular (AV) node*, contractions of the atria followed by relaxations take place. Once the AV node has been signaled, the rapid continuation of the electrical signals occurs through the *AV bundle* and the *right and left bundle branches* within the interventricular septum and terminates via the *Purkinje fibers* throughout the ventricular walls. Once the myocardium of the ventricles has been stimulated, ventricular contractions followed by relaxations occur, completing one cardiac cycle. The sympathetic and para-sympathetic subdivisions of the autonomic nervous system influence the activity of the pacemaker under various conditions. An increased rate results from sympathetic responses; a decreased rate results from parasympathetic responses.

A number of electrical changes also occur in the myocardium as it contracts and relaxes. These changes can be detected by using metal electrodes and an instrument called an *electrocardiograph*. The recording produced by the instrument is an *electrocardiogram,* or *ECG (EKG)*. *Depolarization* and *repolarization* electrical events of the cardiac cycle can be observed and interpreted from the ECG graphic recording. The heart rate can be determined by counting the number of QRS complexes on the ECG during one minute.

Procedure A—Heart Sounds

1. Listen to your heart sounds. To do this, follow these steps:
   a. Obtain a stethoscope and clean its earpieces and the diaphragm by using alcohol swabs.
   b. Fit the earpieces into your ear canals so that the angles are positioned in the forward direction.
   c. Firmly place the diaphragm (or bell) of the stethoscope on the chest over the fifth intercostal space near the apex of the heart (fig. 23.1) and listen to the

FIGURE 23.1   The first sound (lubb) of a cardiac cycle can be heard by placing the diaphragm of a stethoscope over the fifth intercostal space near the apex of the heart. The second sound (dupp) can be heard over the second intercostal space, just left of the sternum. The thoracic regions circled indicate where the sounds of each heart valve are most easily heard.

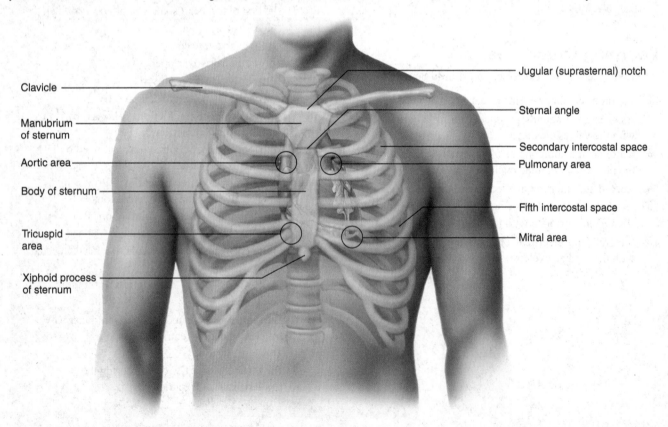

Clavicle

Manubrium of sternum

Aortic area

Body of sternum

Tricuspid area

Xiphoid process of sternum

Jugular (suprasternal) notch

Sternal angle

Secondary intercostal space

Pulmonary area

Fifth intercostal space

Mitral area

sounds. This is a good location to hear the first sound (*lubb*) of a cardiac cycle when the AV valves are closing, which occurs during ventricular *systole* (contraction).

d. Move the diaphragm to the second intercostal space, just to the left of the sternum and listen to the sounds from this region. You should be able to hear the second sound (*dupp*) of the cardiac cycle clearly when the semilunar valves are closing, which occurs during ventricular *diastole* (relaxation).

e. It is possible to hear sounds associated with the aortic and pulmonary valves by listening from the second intercostal space on either side of the sternum. The aortic valve sound comes from the right and the pulmonary valve sound from the left. The sound associated with the mitral valve can be heard from the fifth intercostal space at the nipple line on the left. The sound of the tricuspid valve can be heard at the fifth intercostal space just to the right of the sternum (fig. 23.1).

2. Inhale slowly and deeply and exhale slowly while you listen to the heart sounds from each of the locations as before. Note any changes that have occurred in the sounds.

3. Exercise moderately outside the laboratory for a few minutes so that other students listening to heart sounds will not be disturbed. After the exercise period, listen to the heart sounds and note any changes that have occurred in them. *This exercise should be avoided by anyone with health risks.*

4. Complete Parts A and B of Laboratory Assessment 23.

Procedure B—Electrocardiogram

A single heartbeat is a sequence of precisely timed contractions and relaxations of the four heart chambers initiated from the cardiac conduction system (fig. 23.2). The basic rhythm is controlled from the SA node (pacemaker), but it can be modified by the autonomic nervous system. During atrial contraction (systole), the ventricles are in a state of relaxation (diastole) and are filling with blood forced into them from the atria. Atrial contraction is soon followed by ventricular contraction (systole), during which time atrial relaxation (diastole) occurs. As a result of the synchronized ventricular contractions, blood is forced into the two large arteries that carry blood away from the heart. The aorta receives blood from the left ventricle into the systemic circuit; the pulmonary trunk receives blood from the right ventricle into the pulmonary circuit. This entire sequence of events of a single heartbeat is the cardiac cycle.

Due to the action of the cardiac conduction system, the heart contracts in a coordinated manner. The events of the cardiac conduction system and the cardiac cycle form the basis for the electrocardiogram. The appearance of a normal ECG pattern for one heartbeat is shown in figure 23.3. The ECG components, normal time durations, and corresponding significance of the waves and intervals are

FIGURE 23.2 The cardiac conduction system pathway, indicated by the arrows, from the SA node to the Purkinje fibers (frontal section of heart).

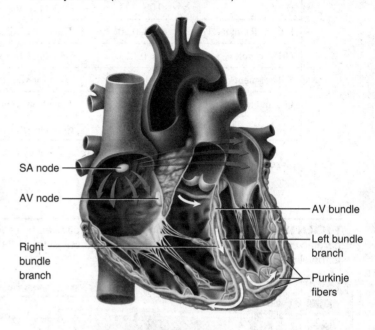

FIGURE 23.3 Components of a normal ECG pattern with a time scale.

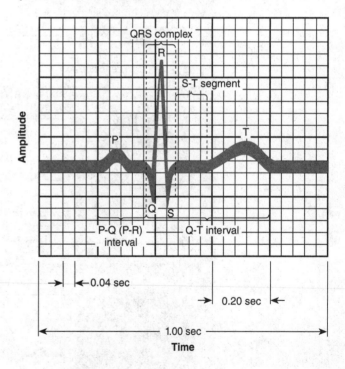

placed in table 23.1. An illustration of the relationship of the ECG components with the depolarization and repolarization of the atria followed by the ventricles is presented in figure 23.4.

1. Study figures 23.2, 23.3, and 23.4, and table 23.1.
2. Complete Part C of the laboratory assessment.

TABLE 23.1 ECG Components, Durations, and Significance

ECG Component	Duration in Seconds	Corresponding Significance
P wave	0.06–0.11	Depolarization of atrial fibers
P-Q (P-R) interval	0.12–0.20	Time for cardiac impulse from SA node through AV node
QRS complex	< 0.12	Depolarization of ventricular fibers
S-T segment	0.12	Time for ventricles to contract
Q-T interval	0.32–0.42	Time from ventricular depolarization to end of ventricular repolarization
T wave	0.16	Repolarization of ventricular fibers (ends pattern)

Note: Atrial repolarization is not observed in the ECG because it occurs at the same time that ventricular fibers depolarize. Therefore it is obscured by the QRS complex.

FIGURE 23.4 Depolarization waves (red) followed by repolarization waves (green) of the atria and ventricles during a cardiac cycle, along with the representative segment of the ECG pattern. The yellow arrows indicate the direction the wave is traveling.

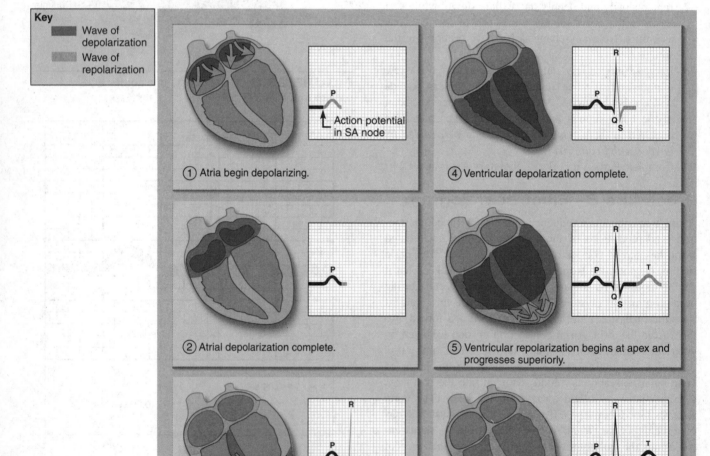

Key
- Wave of depolarization
- Wave of repolarization

① Atria begin depolarizing.

Action potential in SA node

④ Ventricular depolarization complete.

② Atrial depolarization complete.

⑤ Ventricular repolarization begins at apex and progresses superiorly.

③ Ventricular depolarization begins at apex and progresses superiorly as atria repolarize.

⑥ Ventricular repolarization complete; heart is ready for the next cycle.

3. The laboratory instructor will demonstrate the proper adjustment and use of the instrument available to record an electrocardiogram.

4. Record your laboratory partner's ECG. To do this, follow these steps:

   a. Have your partner lie on a cot or table close to the electrocardiograph, remaining as relaxed and still as possible.

   b. Scrub the electrode placement locations with alcohol swabs (fig. 23.5). Apply a small quantity of electrode cream to the skin on the insides of the wrists and ankles. (Any jewelry on the wrists or ankles should be removed.)

   c. Spread some electrode cream over the inner surfaces of four plate electrodes and attach one to each of the prepared skin areas (fig. 23.5). Make sure there is good contact between the skin and the metal of the electrodes. The electrode plate on the right ankle is the grounding system.

   d. Attach the plate electrodes to the corresponding cables of a lead selector switch. When an ECG recording is made, only two electrodes are used at a time, and the selector switch allows various combinations of electrodes (leads) to be activated. Three standard limb leads placed on the two wrists and the left ankle are used for an ECG. This arrangement has become known as *Einthoven's triangle*,[1] which permits the recording of the potential difference between any two of the electrodes (fig. 23.6).

The standard leads I, II, and III are called bipolar leads because they are the potential difference between two electrodes (a positive and a negative). Lead I measures the potential difference between the right wrist (negative) and the left wrist (positive). Lead II measures the potential difference between the right wrist and the left ankle, and lead III measures the potential difference between the left wrist and the left ankle. The right ankle is always the ground.

   e. Turn on the recording instrument and adjust it as previously demonstrated by the laboratory instructor. The paper speed should be set at 2.5 cm/second. This is the standard speed for ECG recordings.

   f. Set the lead selector switch to lead I (right wrist, left wrist electrodes), and record the ECG for 1 minute.

   g. Set the lead selector switch to lead II (right wrist, left ankle electrodes), and record the ECG for 1 minute.

   h. Set the lead selector switch to lead III (left wrist, left ankle electrodes), and record the ECG for 1 minute.

   i. Remove the electrodes and clean the cream from the metal and skin.

   j. Use figure 23.3 to label the ECG components of the results from leads I, II, and III. The P-Q interval is often called the P-R interval because the Q wave is often small or absent. The normal P-Q interval is 0.12–0.20 seconds. The normal QRS complex duration is less than 0.10 seconds.

5. Complete Part D of the laboratory assessment.

---

[1]Willem Einthoven (1860–1927), a Dutch physiologist, received the Nobel prize for physiology or medicine for his work with electrocardiograms.

FIGURE 23.5 To record an ECG, attach electrodes to the wrists and ankles.

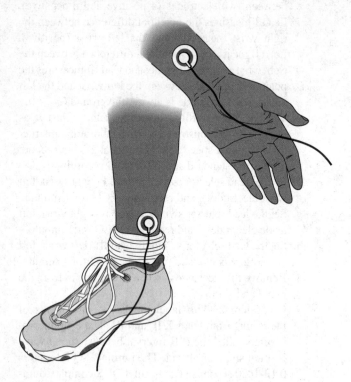

FIGURE 23.6 Standard limb leads for electrocardiograms.

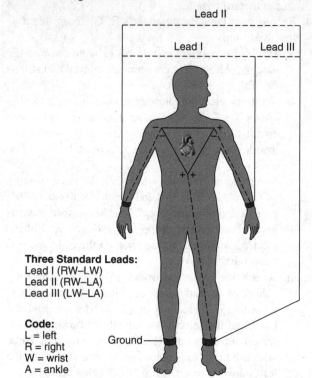

Three Standard Leads:
Lead I (RW–LW)
Lead II (RW–LA)
Lead III (LW–LA)

Code:
L = left
R = right
W = wrist
A = ankle

Name _____

Date _____

Section _____

The ⚠ corresponds to the indicated outcome(s) found at the beginning of the laboratory exercise.

Cardiac Cycle

Part A   Assessments

Complete the following statements:

1. The period during which a heart chamber is contracting is called _____. ⚠

2. The period during which a heart chamber is relaxing is called _____. ⚠

3. During ventricular contraction, the AV valves (tricuspid and mitral valves) are _____. ⚠

4. During ventricular relaxation, the AV valves are _____. ⚠

5. The pulmonary and aortic valves open when the pressure in the _____ exceeds the pressure in the pulmonary trunk and aorta. ⚠

6. The first sound of a cardiac cycle occurs when the _____ are closing. ⚠

7. The second sound of a cardiac cycle occurs when the _____ are closing. ⚠

8. The sound created when blood leaks back through an incompletely closed valve is called a _____. ⚠

Part B   Assessments

Complete the following:

1. What changes did you note in the heart sounds when you inhaled deeply? ⚠

_____

_____

_____

_____

2. What changes did you note in the heart sounds following the exercise period? ⚠

_____

_____

_____

_____

Part C Assessments

Complete the following statements:

1. Normally, the _____ node serves as the pacemaker of the heart. ⓐ

2. The _____ node is located in the inferior portion of the interatrial septum. ⓐ

3. The fibers that carry cardiac impulses from the interventricular septum into the myocardium are called _____ fibers. ⓐ

4. A(n) _____ is a recording of electrical changes occurring in the myocardium during a cardiac cycle. ⓐ

5. The P wave corresponds to depolarization of the muscle fibers of the _____. ⓐ

6. The QRS complex corresponds to depolarization of the muscle fibers of the _____. ⓐ

7. The T wave corresponds to repolarization of the muscle fibers of the _____. ⓐ

8. Why is atrial repolarization not observed in the ECG? ⓐ

_____

_____

Part D Assessments

1. Attach a short segment of the ECG recording from each of the three leads you used, and label the waves of each. Ⓐ

Lead I

Lead II

Lead III

216

2. What differences do you find in the ECG patterns of these leads? A _____

_____

_____

_____

_____

3. How much time passed from the beginning of the P wave to the beginning of the QRS complex (P-Q interval, or P-R interval) in the ECG from lead I? A _____

4. What is the significance of this P-Q (P-R) interval? A _____

_____

_____

_____

_____

5. How can you determine the heart rate from an electrocardiogram? A _____

_____

_____

_____

6. What was your heart rate as determined from the ECG? A _____

🧠 Critical Thinking Assessment

If a person's heart rate is 72 beats per minute, determine the number of QRS complexes that would have appeared on an ECG during the first 30 seconds. A

_____

_____

_____

NOTES

Breathing and Respiratory Volumes

Purpose of the Exercise

To review the mechanisms of breathing and to measure or calculate certain respiratory air volumes and respiratory capacities.

Materials Needed

Lung function model
Spirometer, handheld (dry portable)
Alcohol swabs (wipes)
Disposable mouthpieces
Nose clips
Meterstick

For Learning Extension Activity:
Clock or watch with seconds timer

Safety

▶ Clean the spirometer with an alcohol swab (wipe) before each use.
▶ Place a new disposable mouthpiece on the stem of the spirometer before each use.
▶ Dispose of the alcohol swabs and mouthpieces according to directions from your laboratory instructor.

Learning Outcomes

After completing this exercise, you should be able to

1 Differentiate between the mechanisms responsible for inspiration and expiration.

2 Measure respiratory volumes using a spirometer.

3 Calculate respiratory capacities using the data obtained from respiratory volumes.

4 Match the respiratory volumes and respiratory capacities with their definitions.

5 Explain the changes in tidal volume in response to changes in volume of the anatomic dead space.

6 Apply concepts covered in this laboratory exercise to the changes that occur when the bronchioles are dilated or constricted by the autonomic nervous system.

Pre-Lab

Carefully read the introductory material and examine the entire lab. Be familiar with inspiration, expiration, and respiratory volumes from lecture or the textbook. Answer the pre-lab questions.

Pre-Lab Questions: Select the correct answer for each of the following questions:

1. The size of the thoracic cavity is increased by contractions of all of the following muscles *except* the
   a. diaphragm.   **b.** external intercostals.
   c. pectoralis minor.   **d.** external oblique.

2. A _____ is an instrument to measure air volumes during breathing.
   a. flow meter   **b.** spirometer
   c. lung function model   **d.** capacity meter

3. The _____ is the maximum volume of air that can be exhaled after taking the deepest breath possible.
   a. tidal volume   **b.** expiratory reserve volume
   c. vital capacity   **d.** total lung capacity

4. Tidal volume is estimated to be about
   a. 500 mL.   **b.** 1,200 mL.
   c. 3,600 mL.   **d.** 4,800 mL.

5. A normal resting breathing rate is about _____ breaths per minute.
   a. 5–10   **b.** 12–15
   c. 16–20   **d.** 21–30

6. The contraction of the diaphragm increases the size of the thoracic cavity.
   True _____   False _____

7. Vital capacity is the total of tidal volume, expiratory reserve volume, and residual volume.
   True _____   False _____

8. Vital capacities gradually decrease as a person continues to age.
   True _____   False _____

Breathing, or pulmonary ventilation, involves the movement of air from outside the body through the bronchial tree and into the alveoli and the reversal of this air movement to allow gas (oxygen and carbon dioxide) exchange between air and blood. These movements are caused by changes in the size of the thoracic cavity that result from skeletal muscle contractions. The size of the thoracic cavity during *inspiration* is increased by contractions of the diaphragm, external intercostals, internal intercostals (intercartilaginous part), pectoralis minor, sternocleidomastoid, and scalenes. *Expiration* is aided by contractions of the internal intercostals (interosseous part), rectus abdominis, and external oblique, and from the elastic recoil of stretched tissues (fig. 24.1).

The volumes of air that move in and out of the lungs during various phases of breathing are called *respiratory (pulmonary) volumes* and *capacities*. These volumes can be measured by using an instrument called a *spirometer*. Respiratory capacities can be determined by using various combinations of respiratory volumes. However, the values obtained vary with a person's age, sex, height, weight, stress, and physical fitness.

With each breath you take, what volume of air reaches the alveoli of your lungs? You take a normal breath of air through your nose or mouth. The volume of air you breathe in (or out) is called the *tidal volume.* The average tidal volume is 500 mL (imagine the equivalent of a 1/2 liter bottle). This volume of air is moved in through your nose and mouth and through your respiratory passageway, including the nasal cavity, pharynx, larynx, trachea, and bronchial tree, to the alveoli. A certain amount of that air never reaches your alveoli—it is within the respiratory passageway. This air is said to be in the *anatomic dead space.* The average volume of anatomic dead space is 150 mL. If the tidal volume is 500 mL and the volume of the anatomic dead space is 150 mL, what volume of air reaches your alveoli? Understanding the difference between tidal volume and the volume of air that reaches the lungs is important because *only* the air that reaches the alveoli provides oxygen for gas exchange with the blood.

FIGURE 24.1 Respiratory muscles involved in inspiration and forced expiration. Boldface indicates the primary muscles increasing or decreasing the size of the thoracic cavity. Blue arrows indicate the direction of contraction during inspiration that increases the capacity of the thoracic cavity; the green arrows indicate the direction of contraction during forced expiration that decreases the capacity of the thoracic cavity. (The green arrows associated with the diaphragm indicate its ascending direction primarily from the elastic recoil of the stretched tissues during quiet breathing.)

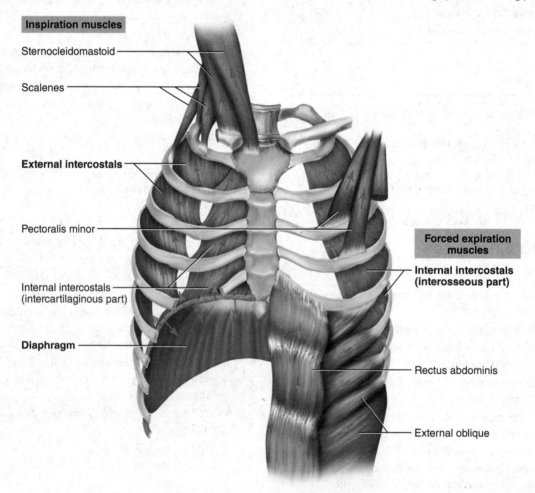

Inspiration muscles

Sternocleidomastoid

Scalenes

External intercostals

Pectoralis minor

Internal intercostals (intercartilaginous part)

Diaphragm

Forced expiration muscles

Internal intercostals (interosseous part)

Rectus abdominis

External oblique

Procedure A—Breathing Mechanisms

The size of the thoracic cavity increases during inspiration by contractions of inspiration muscles, primarily the diaphragm, resulting in increased depth of the thoracic cavity, and external intercostals, widening the thoracic cavity. A combination of synergistic contractions of the sternocleidomastoid, scalenes, pectoralis minor, and the intercartilaginous part of the internal intercostals elevate the sternum and rib cage. During forced expiration, the interosseous part of the internal intercostals depresses the rib cage, narrowing the thoracic cavity. The contractions of the rectus abdominis and the external oblique compress the abdominal viscera, forcing the diaphragm to ascend and thus reducing the size of the thoracic cavity (fig. 24.1).

Air enters the lungs during inspiration and exits the lungs during expiration based upon characteristics of *Boyle's law.* According to Boyle's law, the pressure of a gas is inversely proportional to its volume, assuming a constant temperature. In other words, when the volume increases in a container such as the thoracic cavity during inspiration, pressure decreases and air flows into the lungs. In contrast, when volume decreases in a container such as the thoracic cavity during expiration, pressure increases and air flows out of the lungs. Air moves into and out of the lungs in much the same way as can be demonstrated using a syringe (fig. 24.2). When you pull back on the plunger of the syringe, the volume increases in the barrel of the syringe, which decreases the pressure, resulting in air rushing into the barrel of the syringe. In contrast, pushing the plunger into the barrel of the syringe causes the pressure to increase in the barrel of the syringe, and air flows out of the container.

1. Observe the mechanical lung function model (fig. 24.3). It consists of a heavy plastic bell jar with a rubber sheeting clamped over its wide open end. Its narrow upper opening is plugged with a rubber stopper through which a Y tube is passed. Small rubber balloons are fastened to the arms of the Y. What happens to the balloons when the rubber sheeting is pulled downward?

   _____

   _____

   What happens when the sheeting is pushed upward?

   _____

   _____

2. Compare the lung function model with figures 24.1 and 24.2. Note the muscles of inspiration that increase the size of the thoracic cavity, and the muscles of expiration that decrease the size of the thoracic cavity. Relate these volume changes to pressure changes and airflow into and out of the lungs.

3. Complete Part A of Laboratory Assessment 24.

FIGURE 24.2 Moving the plunger of a syringe as in (a) results in air movements into the barrel of the syringe; moving the plunger of the syringe as in (b) results in air movements out of the barrel of the syringe. Air movements in and out of the lung function model and the lungs during breathing occur for similar reasons.

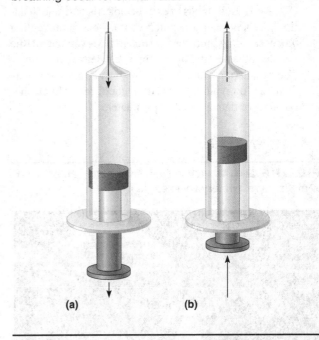

(a)   (b)

FIGURE 24.3 A lung function model.

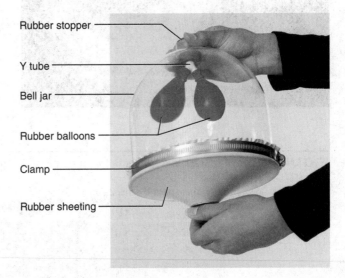

Rubber stopper

Y tube

Bell jar

Rubber balloons

Clamp

Rubber sheeting

Procedure B—Respiratory Volumes and Capacities

 Warning

If the subject begins to feel dizzy or light-headed while performing Procedure B, stop the exercise and breathe normally.

221

1. Get a handheld spirometer (fig. 24.4). Point the needle to zero by rotating the adjustable dial. Before using the instrument, clean it with an alcohol swab and place a new disposable mouthpiece over its stem. The instrument should be held with the dial upward, and *air should be blown only into the disposable mouthpiece* (fig. 24.5). If air tends to exit the nostrils during exhalation, use a nose clip or pinch your nose as a prevention when exhaling into the spirometer. Movement of the needle indicates the air volume that leaves the lungs. The exhalation should be slowly and forcefully performed. Too rapid an exhalation could result in erroneous data or damage to the spirometer.

2. *Tidal volume (TV)* (about 500 mL) is the volume of air that enters (or leaves) the lungs during a *respiratory cycle* (one inspiration plus the following expiration). *Resting tidal volume* is the volume of air that enters (or leaves) the lungs during normal, quiet breathing (fig. 24.6). To measure this volume, follow these steps:
   a. Sit quietly for a few moments.
   b. Position the spirometer dial so that the needle points to zero.

FIGURE 24.5 Demonstration of use of a handheld spirometer. Air should only be blown slowly and forcefully into a disposable mouthpiece. *Use a nose clip or pinch your nose when measuring expiratory reserve volume and vital capacity volume if air exits the nostrils.*

FIGURE 24.4 A handheld spirometer can be used to measure respiratory volumes.

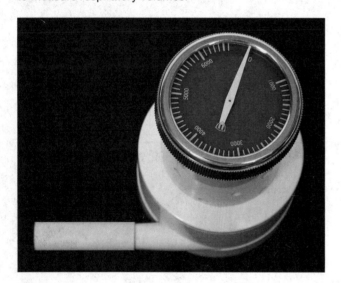

FIGURE 24.6 Graphic representation of respiratory volumes and capacities.

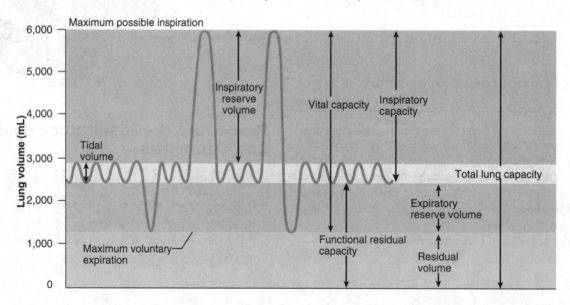

c. Place the mouthpiece between your lips and exhale three ordinary expirations into it after inhaling through the nose each time. *Do not force air out of your lungs; exhale normally.*

d. Divide the total value indicated by the needle by 3 and record this amount as your resting tidal volume on the table in Part B of the laboratory assessment.

3. *Expiratory reserve volume (ERV)* (about 1,200 mL) is the volume of air in addition to the tidal volume that leaves the lungs during forced expiration (fig. 24.6). Chronic obstructive pulmonary disease (COPD) reduces pulmonary ventilation; COPD occurs in emphysema and chronic bronchitis, which are often caused by smoking and other inhaled pollutants. With COPD, the ERV is greatly reduced during physical exertion, and the person becomes easily exhausted from expending more energy just to breathe. To measure this volume, follow these steps:

a. Breathe normally for a few moments. Set the needle to zero.

b. At the end of an ordinary expiration, place the mouthpiece between your lips and exhale all of the air you can force from your lungs through the spirometer. Use a nose clip or pinch your nose to prevent any air from exiting your nostrils.

c. Record the results as your expiratory reserve volume in Part B of the laboratory assessment.

4. *Vital capacity (VC)* (about 4,800 mL) is the maximum volume of air that can be exhaled after taking the deepest breath possible (fig. 24.6). To measure this volume, follow these steps:

a. Breathe normally for a few moments. Set the needle at zero.

b. Breathe in and out deeply a couple of times, then take the deepest breath possible.

c. Place the mouthpiece between your lips and exhale all the air out of your lungs, slowly and forcefully. Use a nose clip or pinch your nose to prevent any air from exiting your nostrils.

d. Record the value as your vital capacity in Part B of the laboratory assessment. Compare your result with that expected for a person of your sex, age, and height listed in tables 24.1 and 24.2. Use the meter-stick to determine your height in centimeters if necessary or multiply your height in inches times 2.54 to calculate your height in centimeters. Considerable individual variations from the expected will be noted due to parameters other than sex, age, and height, which could include physical shape, health, medications, and others.

5. *Inspiratory reserve volume (IRV)* (about 3,100 mL) is the volume of air in addition to the tidal volume that enters the lungs during forced inspiration (fig. 24.6). Calculate your inspiratory reserve volume by subtracting your tidal volume (TV) and your expiratory reserve volume (ERV) from your vital capacity (VC):

$$IRV = VC - (TV + ERV)$$

6. *Inspiratory capacity (IC)* (about 3,600 mL) is the maximum volume of air a person can inhale following exhalation of the tidal volume (fig. 24.6). Calculate your inspiratory capacity by adding your tidal volume (TV) and your inspiratory reserve volume (IRV):

$$IC = TV + IRV$$

7. *Functional residual capacity (FRC)* (about 2,400 mL) is the volume of air that remains in the lungs following exhalation of the tidal volume (fig. 24.6). Calculate your functional residual capacity (FRC) by adding your expiratory reserve volume (ERV) and your residual volume (RV), which you can assume is 1,200 mL:

$$FRC = ERV + 1,200$$

8. *Residual volume (RV)* (about 1,200 mL) is the volume of air that always remains in the lungs after the most forceful expiration (fig. 24.6). Although it is part of *total lung capacity* (about 6,000 mL), it cannot be measured with a spirometer. The residual air allows gas exchange and the alveoli to remain open during the respiratory cycle.

9. Complete Parts B and C of the laboratory assessment.

Learning Extension Activity

Determine your *minute respiratory volume*. To do this, follow these steps:

1. Sit quietly for a while, and then to establish your breathing rate, count the number of times you breathe in 1 minute. This might be inaccurate because conscious awareness of breathing rate can alter the results. You might ask a laboratory partner to record your breathing rate at some time when you are not expecting it to be recorded. A normal resting breathing rate is about 12–15 breaths per minute.

2. Calculate your minute respiratory volume by multiplying your breathing rate by your tidal volume:

| _____ | × | _____ | = | _____ |

(breathing rate)   (tidal volume)   (minute respiratory volume)

3. This value indicates the total volume of air that moves into your respiratory passages during each minute of ordinary breathing.

TABLE 24.1 Predicted Vital Capacities (in Milliliters) for Females

Age	Height in Centimeters																								
	146	148	150	152	154	156	158	160	162	164	166	168	170	172	174	176	178	180	182	184	186	188	190	192	194
16	2950	2990	3030	3070	3110	3150	3190	3230	3270	3310	3350	3390	3430	3470	3510	3550	3590	3630	3670	3715	3755	3800	3840	3880	3920
18	2920	2960	3000	3040	3080	3120	3160	3200	3240	3280	3320	3360	3400	3440	3480	3520	3560	3600	3640	3680	3720	3760	3800	3840	3880
20	2890	2930	2970	3010	3050	3090	3130	3170	3210	3250	3290	3330	3370	3410	3450	3490	3525	3565	3605	3645	3695	3720	3760	3800	3840
22	2860	2900	2940	2980	3020	3060	3095	3135	3175	3215	3255	3290	3330	3370	3410	3450	3490	3530	3570	3610	3650	3685	3725	3765	3800
24	2830	2870	2910	2950	2985	3025	3065	3100	3140	3180	3220	3260	3300	3335	3375	3415	3455	3490	3530	3570	3610	3650	3685	3725	3765
26	2800	2840	2880	2920	2960	3000	3035	3070	3110	3150	3190	3230	3265	3300	3340	3380	3420	3455	3495	3530	3570	3610	3650	3685	3725
28	2775	2810	2850	2890	2930	2965	3000	3040	3070	3115	3155	3190	3230	3270	3305	3345	3380	3420	3460	3495	3535	3570	3610	3650	3685
30	2745	2780	2820	2860	2895	2935	2970	3010	3045	3085	3120	3160	3195	3235	3270	3310	3345	3385	3420	3460	3495	3535	3570	3610	3645
32	2715	2750	2790	2825	2865	2900	2940	2975	3015	3050	3090	3125	3160	3200	3235	3275	3310	3350	3385	3425	3460	3495	3535	3570	3610
34	2685	2725	2760	2795	2835	2870	2910	2945	2980	3020	3055	3090	3130	3165	3200	3240	3275	3310	3350	3385	3425	3460	3495	3535	3570
36	2655	2695	2730	2765	2805	2840	2875	2910	2950	2985	3020	3060	3095	3130	3165	3205	3240	3275	3310	3350	3385	3420	3460	3495	3530
38	2630	2665	2700	2735	2770	2810	2845	2880	2915	2950	2990	3025	3060	3095	3130	3170	3205	3240	3275	3310	3350	3385	3420	3455	3490
40	2600	2635	2670	2705	2740	2775	2810	2850	2885	2920	2955	2990	3025	3060	3095	3135	3170	3205	3240	3275	3310	3345	3380	3420	3455
42	2570	2605	2640	2675	2710	2745	2780	2815	2850	2885	2920	2955	2990	3025	3060	3100	3135	3170	3205	3240	3275	3310	3345	3380	3415
44	2540	2575	2610	2645	2680	2715	2750	2785	2820	2855	2890	2925	2960	2995	3030	3060	3095	3130	3165	3200	3235	3270	3305	3340	3375
46	2510	2545	2580	2615	2650	2685	2715	2750	2785	2820	2855	2890	2925	2960	2995	3030	3060	3095	3130	3165	3200	3235	3270	3305	3340
48	2480	2515	2550	2585	2620	2650	2685	2715	2750	2785	2820	2855	2890	2925	2960	2995	3030	3060	3095	3130	3160	3195	3230	3265	3300
50	2455	2485	2520	2555	2590	2625	2655	2690	2715	2750	2785	2820	2855	2890	2925	2955	2990	3025	3060	3090	3125	3155	3190	3225	3260
52	2425	2455	2490	2525	2555	2590	2625	2655	2690	2720	2755	2790	2820	2855	2890	2925	2955	2990	3020	3055	3090	3125	3155	3190	3220
54	2395	2425	2460	2495	2530	2560	2590	2625	2655	2690	2720	2755	2790	2820	2855	2885	2920	2950	2985	3020	3050	3085	3115	3150	3180
56	2365	2400	2430	2460	2495	2525	2560	2590	2625	2655	2690	2720	2755	2790	2820	2855	2885	2920	2950	2980	3015	3045	3080	3110	3145
58	2335	2370	2400	2430	2460	2495	2525	2560	2590	2625	2655	2690	2720	2750	2785	2815	2850	2880	2920	2945	2975	3010	3040	3075	3105
60	2305	2340	2370	2400	2430	2460	2495	2525	2560	2590	2625	2655	2685	2720	2750	2780	2810	2845	2875	2915	2940	2970	3000	3035	3065
62	2280	2310	2340	2370	2405	2435	2465	2495	2525	2560	2590	2620	2655	2685	2715	2745	2775	2810	2840	2870	2900	2935	2965	2995	3025
64	2250	2280	2310	2340	2370	2400	2430	2465	2495	2525	2555	2585	2620	2650	2680	2710	2740	2770	2805	2835	2865	2895	2920	2955	2990
66	2220	2250	2280	2310	2340	2370	2400	2430	2460	2495	2525	2555	2585	2615	2645	2675	2705	2735	2765	2800	2825	2860	2890	2920	2950
68	2190	2220	2250	2280	2310	2340	2370	2400	2430	2460	2490	2520	2550	2580	2610	2640	2670	2700	2730	2760	2795	2820	2850	2880	2910
70	2160	2190	2220	2250	2280	2310	2340	2370	2400	2425	2455	2485	2515	2545	2575	2605	2635	2665	2695	2725	2755	2780	2810	2840	2870
72	2130	2160	2190	2220	2250	2280	2310	2335	2365	2395	2425	2455	2480	2510	2540	2570	2600	2630	2660	2685	2715	2745	2775	2805	2830
74	2100	2130	2160	2190	2220	2245	2275	2305	2335	2360	2390	2420	2450	2475	2505	2535	2565	2590	2620	2650	2680	2710	2740	2765	2795

From E. DeF. Baldwin and E. W. Richards, Jr., "Pulmonary Insufficiency 1, Physiologic Classification, Clinical Methods of Analysis, Standard Values in Normal Subjects" in *Medicine* 27:243, © by William & Wilkins. Used by permission.

TABLE 24.2 Predicted Vital Capacities (in Milliliters) for Males

Age	\multicolumn{25}{c}{Height in Centimeters}
	146	148	150	152	154	156	158	160	162	164	166	168	170	172	174	176	178	180	182	184	186	188	190	192	194
16	3765	3820	3870	3920	3975	4025	4075	4130	4180	4230	4285	4335	4385	4440	4490	4540	4590	4645	4695	4745	4800	4850	4900	4955	5005
18	3740	3790	3840	3890	3940	3995	4045	4095	4145	4200	4250	4300	4350	4405	4455	4505	4555	4610	4660	4710	4760	4815	4865	4915	4965
20	3710	3760	3810	3860	3910	3960	4015	4065	4115	4165	4215	4265	4320	4370	4420	4470	4520	4570	4625	4675	4725	4775	4825	4875	4930
22	3680	3730	3780	3830	3880	3930	3980	4030	4080	4135	4185	4235	4285	4335	4385	4435	4485	4535	4585	4635	4685	4735	4790	4840	4890
24	3635	3685	3735	3785	3835	3885	3935	3985	4035	4085	4135	4185	4235	4285	4330	4380	4430	4480	4530	4580	4630	4680	4730	4780	4830
26	3605	3655	3705	3755	3805	3855	3905	3955	4000	4050	4100	4150	4200	4250	4300	4350	4395	4445	4495	4545	4595	4645	4695	4740	4790
28	3575	3625	3675	3725	3775	3820	3870	3920	3970	4020	4070	4115	4165	4215	4265	4310	4360	4410	4460	4510	4555	4605	4655	4705	4755
30	3550	3595	3645	3695	3740	3790	3840	3890	3935	3985	4035	4080	4130	4180	4230	4275	4325	4375	4425	4470	4520	4570	4615	4665	4715
32	3520	3565	3615	3665	3710	3760	3810	3855	3905	3950	4000	4050	4095	4145	4195	4240	4290	4340	4385	4435	4485	4530	4580	4625	4675
34	3475	3525	3570	3620	3665	3715	3760	3810	3855	3905	3950	4000	4045	4095	4140	4190	4225	4285	4330	4380	4425	4475	4520	4570	4615
36	3445	3495	3540	3585	3635	3680	3730	3775	3825	3870	3920	3965	4010	4060	4105	4155	4200	4250	4295	4340	4390	4435	4485	4530	4580
38	3415	3465	3510	3555	3605	3650	3695	3745	3790	3840	3885	3930	3980	4025	4070	4120	4165	4210	4260	4305	4350	4400	4445	4495	4540
40	3385	3435	3480	3525	3575	3620	3665	3710	3760	3805	3850	3900	3945	3990	4035	4085	4130	4175	4220	4270	4315	4360	4410	4455	4500
42	3360	3405	3450	3495	3540	3590	3635	3680	3725	3770	3820	3865	3910	3955	4000	4050	4095	4140	4185	4230	4280	4325	4370	4415	4460
44	3315	3360	3405	3450	3495	3540	3585	3630	3675	3725	3770	3815	3860	3905	3950	3995	4040	4085	4130	4175	4220	4270	4315	4360	4405
46	3285	3330	3375	3420	3465	3510	3555	3600	3645	3690	3735	3780	3825	3870	3915	3960	4005	4050	4095	4140	4185	4230	4275	4320	4365
48	3255	3300	3345	3390	3435	3480	3525	3570	3615	3655	3700	3745	3790	3835	3880	3925	3970	4015	4060	4105	4150	4190	4235	4280	4325
50	3210	3255	3300	3345	3390	3430	3475	3520	3565	3610	3650	3695	3740	3785	3830	3870	3915	3960	4005	4050	4090	4135	4180	4225	4270
52	3185	3225	3270	3315	3355	3400	3445	3490	3530	3575	3620	3660	3705	3750	3795	3835	3880	3925	3970	4010	4055	4100	4140	4185	4230
54	3155	3195	3240	3285	3325	3370	3415	3455	3500	3540	3585	3630	3670	3715	3760	3800	3845	3890	3930	3975	4020	4060	4105	4145	4190
56	3125	3165	3210	3255	3295	3340	3380	3425	3465	3510	3550	3595	3640	3680	3725	3765	3810	3850	3895	3940	3980	4025	4065	4110	4150
58	3080	3125	3165	3210	3250	3290	3335	3375	3420	3460	3500	3545	3585	3630	3670	3715	3755	3800	3840	3880	3925	3965	4010	4050	4095
60	3050	3095	3135	3175	3220	3260	3300	3345	3385	3430	3470	3500	3555	3595	3635	3680	3720	3760	3805	3845	3885	3930	3970	4015	4055
62	3020	3060	3110	3150	3190	3230	3270	3310	3350	3390	3440	3480	3520	3560	3600	3640	3680	3730	3770	3810	3850	3890	3930	3970	4020
64	2990	3030	3080	3120	3160	3200	3240	3280	3320	3360	3400	3440	3490	3530	3570	3610	3650	3690	3730	3770	3810	3850	3900	3940	3980
66	2950	2990	3030	3070	3110	3150	3190	3230	3270	3310	3350	3390	3430	3470	3510	3550	3600	3640	3680	3720	3760	3800	3840	3880	3920
68	2920	2960	3000	3040	3080	3120	3160	3200	3240	3280	3320	3360	3400	3440	3480	3520	3560	3600	3640	3680	3720	3760	3800	3840	3880
70	2890	2930	2970	3010	3050	3090	3130	3170	3210	3250	3290	3330	3370	3410	3450	3480	3520	3560	3600	3640	3680	3720	3760	3800	3840
72	2860	2900	2940	2980	3020	3060	3100	3140	3180	3210	3250	3290	3330	3370	3410	3450	3490	3530	3570	3610	3650	3680	3720	3760	3800
74	2820	2860	2900	2930	2970	3010	3050	3090	3130	3170	3200	3240	3280	3320	3360	3400	3440	3470	3510	3550	3590	3630	3670	3710	3740

From E. DeF. Baldwin and E. W. Richards, Jr., "Pulmonary Insufficiency 1, Physiologic Classification, Clinical Methods of Analysis, Standard Values in Normal Subjects" in *Medicine* 27:243, © by William & Wilkins. Used by permission.

Name _____

Date _____

Section _____

The Ⓐ corresponds to the indicated outcome(s) found at the beginning of the laboratory exercise.

Breathing and Respiratory Volumes

Part A Assessments

Complete the following statements:

1. When using the lung function model, what part of the respiratory system is represented by: Ⓐ

   a. The rubber sheeting? _____   **c.** The Y tube? _____

   b. The bell jar? _____   **d.** The balloons? _____

2. When the diaphragm contracts, the size of the thoracic cavity _____. Ⓐ

3. The ribs are raised by contraction of the _____

   muscles, which increases the size of the thoracic cavity. Ⓐ

4. Muscles that help to force out more than the normal volume of air by pulling the ribs downward and inward include the

   _____. Ⓐ

5. We inhale when the diaphragm _____. Ⓐ

Part B Assessments

1. Test results for respiratory air volumes and capacities: Ⓐ Ⓐ

Respiratory Volume or Capacity	Expected Value* (approximate)	Test Result	Percent of Expected Value (test result/expected value × 100)
Tidal volume (resting) (TV)	500 mL		
Expiratory reserve volume (ERV)	1,200 mL		
Vital capacity (VC)	(enter yours from table 24.1 or 24.2)		
Inspiratory reserve volume (IRV)	3,100 mL		
Inspiratory capacity (IC)	3,600 mL		
Functional residual capacity (FRC)	2,400 mL		

*The values listed are most characteristic for a healthy, tall, young adult male. In general, adult females have smaller bodies and therefore smaller lung volumes and capacities. If your expected value for vital capacity is considerably different than 4,800 mL, your other values will vary accordingly.

2. Complete the following:

    a. How do your test results compare with the expected values? _____

    _____

    _____

    _____

    b. How does your vital capacity compare with the average value for a person of your sex, age, and height?

    _____

    _____

    c. What measurement in addition to vital capacity is needed before you can calculate your total lung capacity?

    _____

3. If your experimental results are considerably different than the predicted vital capacities, propose reasons for the differences. As you write this paragraph, consider factors such as smoking, physical fitness, respiratory disorders, stress, and medications. (Your instructor might have you make some class correlations from class data.)

_____

_____

_____

_____

Part C Assessments

Match the air volumes in column A with the definitions in column B. Place the letter of your choice in the space provided. ◢

Column A	Column B
a. Expiratory reserve volume	_____ **1.** Volume in addition to tidal volume that leaves the lungs during forced expiration
b. Functional residual capacity	
c. Inspiratory capacity	_____ **2.** Vital capacity plus residual volume
d. Inspiratory reserve volume	_____ **3.** Volume that remains in lungs after the most forceful expiration
e. Residual volume	
f. Tidal volume	_____ **4.** Volume that enters or leaves lungs during a respiratory cycle
g. Total lung capacity	_____ **5.** Volume in addition to tidal volume that enters lungs during forced inspiration
h. Vital capacity	
	_____ **6.** Maximum volume a person can exhale after taking the deepest possible breath
	_____ **7.** Maximum volume a person can inhale following exhalation of the tidal volume
	_____ **8.** Volume of air remaining in the lungs following exhalation of the tidal volume

Critical Thinking Assessment 6

If the bronchioles are dilated (when the sympathetic nervous system is activated, for example), the anatomic dead space _____ (increases, decreases, or remains the same), and tidal volume will _____ (increase, decrease, or remain the same). In patients with emphysema, the anatomic dead space increases (since there is a loss of elasticity of lung tissue and the thoracic cage can not effectively "be pulled back in"). Predict the changes in breathing of a patient with emphysema as he/she tries to compensate for the increase in anatomic dead space. _____

_____

_____

_____

_____

Control of Breathing

Purpose of the Exercise

To review the muscles and the mechanisms that control breathing and to investigate some of the factors that affect the rate and depth of breathing.

Materials Needed

Clock or watch with seconds timer
Paper bags, small

For Demonstration Activity:

Flasks
Glass tubing
Rubber stoppers, two-hole
Calcium hydroxide solution (limewater)

For Learning Extension Activity:

Pneumograph
Physiological recording apparatus

Learning Outcomes

After completing this exercise, you should be able to

1. Locate the respiratory areas in the brainstem.
2. Describe the mechanisms that control and influence breathing.
3. Select the respiratory muscles involved in inspiration and forced expiration.
4. Test and record the effect of various factors on the rate and depth of breathing.

Pre-Lab

Carefully read the introductory material and examine the entire lab. Be familiar with control of breathing from lecture or the textbook. Answer the pre-lab questions.

Pre-Lab Questions: Select the correct answer for each of the following questions:

1. Respiratory areas of the brain include all of the following *except* the
   a. brainstem.   b. pons.
   c. medulla oblongata. d. pineal gland.

2. Breathing rate increases as blood concentrations of
   a. carbon dioxide increase.
   b. carbon dioxide decrease.
   c. hydrogen ions decrease.
   d. oxygen increase.

3. Forced expiration muscles do *not* include the
   a. internal intercostal muscles.
   b. rectus abdominis muscles.
   c. sternocleidomastoid muscles.
   d. external oblique muscles.

4. Peripheral chemoreceptors sensitive to low blood oxygen levels are located in the
   a. heart.
   b. aortic arch and carotid arteries.
   c. aorta only.
   d. carotid arteries only.

5. Normal blood pH is
   a. 6.8.   b. 8.0.
   c. 6.8–8.0.   d. 7.35–7.45.

6. An increase in the duration of inspirations is the normal response from the inflation reflex.
   True _____   False _____

7. The dorsal respiratory group of the medullary respiratory center is involved with stimulation of the diaphragm contractions.
   True _____   False _____

Breathing is controlled from regions of the brainstem called the *respiratory areas,* which control both inspiration and expiration. These areas initiate and regulate nerve impulses that travel to various breathing muscles, causing rhythmic breathing movements and adjustments to the rate and depth of breathing to meet various cellular needs (fig. 25.1). The *medullary respiratory center* is composed of two bilateral groups of neurons in the medulla oblongata. They are called the *dorsal respiratory group* and the *ventral respiratory group.* The dorsal respiratory group primarily stimulates the diaphragm to contract, resulting in inspiration, and

helps process respiratory sensory information involving the cardiovascular system. The ventral respiratory group involves regulation of the basic rhythm of breathing. Neurons in another part of the brainstem, the pons, compose the *pontine respiratory group* (formerly the pneumotaxic center). These neurons may contribute to the rhythm of breathing by modifying the respirations during situations such as exercise and sleep (fig. 25.2).

Various factors can influence the respiratory areas and thus affect the rate and depth of breathing. These factors include stretch of the lung tissues, emotional state, and the presence in the blood of certain chemicals, such as carbon dioxide, hydrogen ions, and oxygen. The breathing rate increases as the blood concentration of carbon dioxide or hydrogen ions increases or as the concentration of oxygen decreases.

The medulla oblongata possesses *central chemoreceptors* that are sensitive to changes in the levels of carbon dioxide and hydrogen ions. If the levels of carbon dioxide or hydrogen ions increase, the chemoreceptors relay this information to the respiratory areas. As a result, the rate and depth of breathing increase and more carbon dioxide is exhaled. When the levels decline to a more normal range, breathing rate decreases to normal. Exercise, breath holding, and hyperventilation are examples that considerably alter the blood levels of carbon dioxide and hydrogen ions.

Low blood oxygen levels are detected by *peripheral chemoreceptors* located in *carotid bodies* in the carotid arteries and *aortic bodies* in the aortic arch (fig. 25.3). This information is relayed to the respiratory areas in the brainstem, and the breathing rate increases. In order to trigger this response, the blood oxygen levels must be very low. Therefore, the blood levels of carbon dioxide and hydrogen ions have a much greater influence upon the rate and depth of breathing.

The depth of breathing is also influenced by the *inflation reflex.* If the lungs are greatly expanded during forceful

FIGURE 25.1 The respiratory areas are located in the pons and the medulla oblongata.

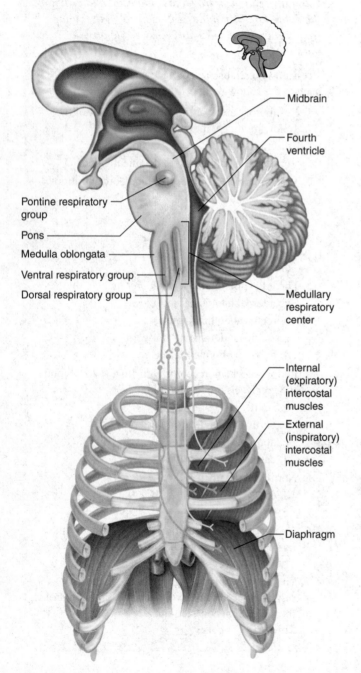

- Midbrain
- Fourth ventricle
- Pontine respiratory group
- Pons
- Medulla oblongata
- Ventral respiratory group
- Dorsal respiratory group
- Medullary respiratory center
- Internal (expiratory) intercostal muscles
- External (inspiratory) intercostal muscles
- Diaphragm

FIGURE 25.2 The medullary respiratory center and the pontine respiratory group control breathing.

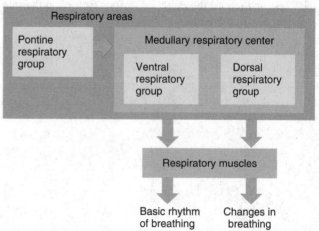

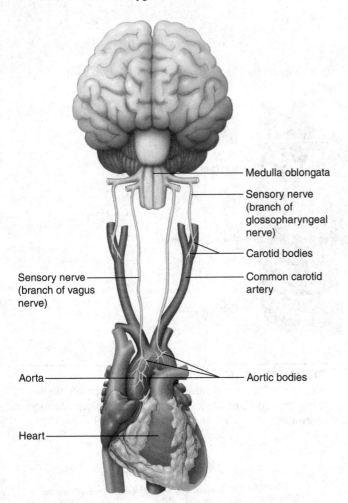

- Medulla oblongata
- Sensory nerve (branch of glossopharyngeal nerve)
- Carotid bodies
- Common carotid artery
- Sensory nerve (branch of vagus nerve)
- Aorta
- Aortic bodies
- Heart

breathing, stretch receptors within the bronchial tree and the visceral pleura are stimulated. Sensory impulses are conducted to the pontine respiratory group and, as a result, a decrease occurs in the duration of inspirations.

The muscles involved with inspiration and expiration are all voluntary skeletal muscles. Therefore a person has some cerebral conscious influence over the rate and depth of breathing as in talking, singing, and breath holding. However, respiratory areas of the brainstem, chemoreceptors, and certain reflexes represent the primary mechanism of subconscious, automatic control of breathing. The complex coordination of breathing involves considerable integration between respiratory areas of the brain and receptors; however some details of the control of breathing are still obscure.

Procedure A—Control of Breathing

1. Several skeletal muscles contract during breathing. The principal muscles involved during inspiration are the diaphragm and external intercostals. The sternocleido-

mastoid, scalenes, and pectoralis minor are synergistic during more forceful inhalation, resulting in greater volumes of air inhaled. During quiet respiration, there is minimal involvement of the expiratory muscles. During forced expiration, the principal muscles are the internal intercostals, but the rectus abdominis and the external oblique muscles can provide extra force (fig. 25.4).

2. Respiratory areas in the brainstem control the cycle of breathing. The dorsal respiratory group of neurons is located in the medulla oblongata. Impulses from the dorsal respiratory group stimulate the muscles of inspiration, especially the diaphragm (fig. 25.1). This results in a normal respiratory rate of 12 to 15 breaths per minute. The ventral respiratory group of neurons will fire during inspiration and expiration to control the appropriate muscles of inspiration and expiration and the basic rhythm of breathing (fig. 25.2).

3. Complete Part A of Laboratory Assessment 25.

Procedure B—Factors Affecting Breathing

Perform each of the following tests, using your laboratory partner as a test subject.

1. *Normal breathing.* To determine the subject's normal breathing rate and depth, follow these steps:
   a. Have the subject sit quietly for a few minutes.
   b. After the rest period, ask the subject to count backwards mentally, beginning with five hundred.
   c. While the subject is distracted by counting, watch the subject's chest movements, and count the breaths taken in a minute. Use this value as the normal breathing rate (breaths per minute).
   d. Note the relative depth of the breathing movements.
   e. Record your observations in the table in Part B of the laboratory assessment.

2. *Effect of hyperventilation.* To test the effect of hyperventilation on breathing, follow these steps:
   a. Seat the subject and *guard to prevent the possibility of the subject falling over.*
   b. Have the subject breathe rapidly and deeply for a maximum of 20 seconds. *If the subject begins to feel dizzy, the hyperventilation should be halted immediately to prevent the subject from fainting from complications of alkalosis. The increased blood pH causes vasoconstriction of cerebral arterioles, which decreases circulation and oxygen to the brain.*
   c. After the period of hyperventilation, determine the subject's breathing rate and judge the breathing depth as before.
   d. Record the results in Part B of the laboratory assessment.

3. *Effect of rebreathing air.* To test the effect of rebreathing air on breathing, follow these steps:
   a. Have the subject sit quietly (approximately 5 minutes) until the breathing rate returns to normal.

FIGURE 25.4 Respiratory muscles involved in inspiration and forced expiration. The blue arrows indicate the direction of muscle contraction during inspiration; the green arrows indicate the direction of muscle contraction during forced expiration. Boldface indicates the primary muscles involved in breathing.

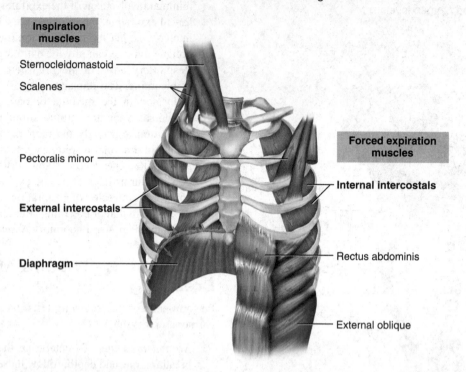

Inspiration muscles
- Sternocleidomastoid
- Scalenes
- Pectoralis minor
- **External intercostals**
- **Diaphragm**

Forced expiration muscles
- **Internal intercostals**
- Rectus abdominis
- External oblique

Demonstration Activity

When a solution of calcium hydroxide is exposed to carbon dioxide, a chemical reaction occurs, and a white precipitate of calcium carbonate is formed as indicated by the following reaction:

$$Ca(OH)_2 + CO_2 \longrightarrow CaCO_3 + H_2O$$

Thus, a clear water solution of calcium hydroxide (limewater) can be used to detect the presence of carbon dioxide because the solution becomes cloudy if this gas is bubbled through it.

The laboratory instructor will demonstrate this test for carbon dioxide by drawing some air through limewater in an apparatus such as that shown in figure 25.5. Then the instructor will blow an equal volume of expired air through a similar apparatus. (*Note:* A new sterile mouthpiece should be used each time the apparatus is demonstrated.) Watch for the appearance of a precipitate that causes the limewater to become cloudy. Was there any carbon dioxide in the atmospheric air drawn through the limewater?

_____

If so, how did the amount of carbon dioxide in the atmospheric air compare with the amount in the expired air?

_____

_____

FIGURE 25.5 Apparatus used to demonstrate the presence of carbon dioxide in air: (a) atmospheric air is drawn through limewater; (b) expired air is blown through limewater.

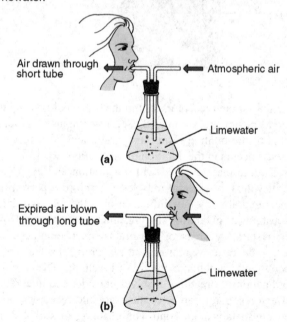

Air drawn through short tube ← Atmospheric air

Limewater

(a)

Expired air blown through long tube

Limewater

(b)

b. Have the subject breathe deeply into a small paper bag that is held tightly over the nose and mouth. *If the subject begins to feel light-headed or like fainting, the rebreathing air should be halted immediately to prevent further acidosis and fainting.*

c. After 2 minutes of rebreathing air, determine the subject's breathing rate and judge the depth of breathing.

d. Record the results in Part B of the laboratory assessment.

4. *Effect of breath holding.* To test the effect of breath holding on breathing, follow these steps:

a. Have the subject sit quietly (approximately 5 minutes) until the breathing rate returns to normal.

b. Have the subject hold his or her breath as long as possible. *If the subject begins to feel light-headed or like fainting, breath holding should be halted immediately to prevent further acidosis and fainting.*

c. As the subject begins to breathe again, determine the rate of breathing and judge the depth of breathing.

d. Record the results in Part B of the laboratory assessment.

5. *Effect of exercise.* To test the effect of exercise on breathing, follow these steps:

a. Have the subject sit quietly (approximately 5 minutes) until breathing rate returns to normal.

b. Have the subject exercise by moderately running in place for 3-5 minutes. *This exercise should be avoided by anyone with health risks.*

c. After the exercise, determine the breathing rate and judge the depth of breathing.

d. Record the results in Part B of the laboratory assessment.

6. The importance of the respiratory rate and blood pH relationship is illustrated in figure 25.6. A condition of respiratory acidosis or alkalosis, leading to possible death, can occur from changes in the depth or rate of breathing.

7. Complete Part B of the laboratory assessment.

FIGURE 25.6 The relationship of respiratory rate and blood pH. The normal breathing rate of 12–15 breaths per minute relates to a blood pH within the normal range.

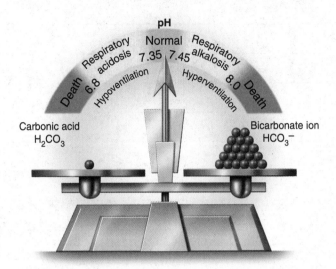

Learning Extension Activity

A *pneumograph* is a device that can be used together with some type of recording apparatus to record breathing movements. The laboratory instructor will demonstrate the use of this equipment to record various movements, such as those that accompany coughing, laughing, yawning, and speaking.

Devise an experiment to test the effect of some factor, such as hyperventilation, rebreathing air, or exercise, on the length of time a person can hold the breath. *After the laboratory instructor has approved your plan,* carry out the experiment, using the pneumograph and recording equipment. What conclusion can you draw from the results of your experiment?

_____

_____

_____

_____

Name _____

Date _____

Section _____

The ◬ corresponds to the indicated outcome(s) found at the beginning of the laboratory exercise.

Control of Breathing

Part A Assessments

Complete the following statements:

1. The respiratory areas are widely scattered throughout the _____ and medulla oblongata of the brainstem. ◬

2. The _____ respiratory group within the medulla oblongata primarily stimulates the diaphragm. ◬

3. The _____ respiratory group within the medulla oblongata regulates the basic rhythm of breathing. ◬

4. Central chemoreceptors are sensitive to changes in the blood concentrations of hydrogen ions and _____. ◬

5. As the blood concentration of carbon dioxide increases, the breathing rate _____. ◬

6. As a result of increased breathing, the blood concentration of carbon dioxide is _____. ◬

7. Peripheral chemoreceptors include aortic bodies and the _____. ◬

8. The principal muscles of forced expiration are the _____. ◬

9. The principal muscles of inspiration are the _____ and the external intercostal muscles. ◬

Part B Assessments

1. Record the results of your breathing tests in the table. ◬

Factor Tested	Breathing Rate (breaths/minute)	Breathing Depth (+, + +, + + +)
Normal		
Hyperventilation (shortly after)		
Rebreathing air (shortly after)		
Breath holding (shortly after)		
Exercise (shortly after)		

2. Briefly explain the reason for the changes in breathing that occurred in each of the following cases: 🔼

   a. Hyperventilation

_____

_____

_____

_____

   b. Rebreathing air

_____

_____

_____

_____

   c. Breath holding

_____

_____

_____

_____

   d. Exercise

_____

_____

_____

_____

3. Complete the following:

   a. Why is it important to distract a person when you are determining the normal rate of breathing?

_____

_____

   b. How can the depth of breathing be measured accurately?

_____

_____

Critical Thinking Assessment

Why is it dangerous for a swimmer to hyperventilate in order to hold the breath for a longer period of time? 🔼

_____

_____

_____

_____

Digestive Organs

Purpose of the Exercise

To review the structure and function of the digestive organs and to examine the tissues of these organs.

Materials Needed

Human torso model
Head model, sagittal section
Skull with teeth
Teeth, sectioned
Tooth model, sectioned
Paper cup
Compound light microscope
Prepared microscope slides of the following:
  Sublingual gland (salivary gland)
  Esophagus
  Stomach (fundus)
  Pancreas (exocrine portion)
  Small intestine (jejunum)
  Large intestine

Learning Outcomes

After completing this exercise, you should be able to

1. Sketch and label the structures of tissue sections of digestive organs.
2. Locate the major organs and structural features of the digestive system.
3. Match digestive organs with their descriptions.
4. Describe the functions of these organs.

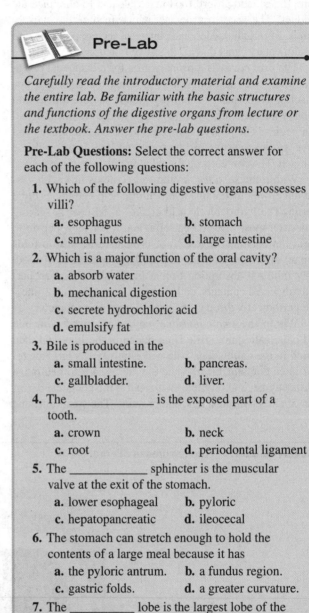

Pre-Lab

Carefully read the introductory material and examine the entire lab. Be familiar with the basic structures and functions of the digestive organs from lecture or the textbook. Answer the pre-lab questions.

Pre-Lab Questions: Select the correct answer for each of the following questions:

1. Which of the following digestive organs possesses villi?
   - a. esophagus
   - b. stomach
   - c. small intestine
   - d. large intestine

2. Which is a major function of the oral cavity?
   - a. absorb water
   - b. mechanical digestion
   - c. secrete hydrochloric acid
   - d. emulsify fat

3. Bile is produced in the
   - a. small intestine.
   - b. pancreas.
   - c. gallbladder.
   - d. liver.

4. The _____ is the exposed part of a tooth.
   - a. crown
   - b. neck
   - c. root
   - d. periodontal ligament

5. The _____ sphincter is the muscular valve at the exit of the stomach.
   - a. lower esophageal
   - b. pyloric
   - c. hepatopancreatic
   - d. ileocecal

6. The stomach can stretch enough to hold the contents of a large meal because it has
   - a. the pyloric antrum.
   - b. a fundus region.
   - c. gastric folds.
   - d. a greater curvature.

7. The _____ lobe is the largest lobe of the liver.
   - a. left
   - b. quadrate
   - c. caudate
   - d. right

8. The appendix, cecum, and ascending colon are on the left side of the body.
   True _____ False _____

9. The anal sphincter muscles include an external voluntary sphincter and an internal involuntary sphincter.
   True _____ False _____

The digestive system includes the organs associated with the *alimentary canal* and several accessory structures. The alimentary canal, a muscular tube, passes through the body from the opening of the mouth to the anus. It includes the mouth, pharynx, esophagus, stomach, small intestine, and large intestine. The canal is adapted to move substances throughout its length. It is specialized in various regions to store, digest, and absorb food materials and to eliminate the residues. The accessory organs, which include the salivary glands, liver, gallbladder, and pancreas, secrete products into the alimentary canal that aid digestive functions.

The *oral cavity* is a primary area for mechanical digestion where we masticate the food using the teeth and jaw muscles. Three pairs of salivary glands secrete mucus and salivary amylase into the oral cavity, where chemical digestion begins. The *pharynx* and *esophagus* secrete mucus and serve as passageways for the food and liquids to reach the stomach. The contents are forced along by peristaltic waves.

Within the *stomach,* swallowed contents mix with gastric juice, which contains hydrochloric acid and the enzyme pepsin. The hydrochloric acid creates a pH near 2, serving to destroy most ingested bacteria, and it activates pepsin to begin the chemical digestion of proteins. The gastric folds (rugae) allow the stomach to hold large quantities of food. Only minimal absorption occurs in the stomach. The partially digested contents, called chyme, enter the small intestine periodically through the muscular pyloric sphincter.

Within the *small intestine,* chyme is mixed with bile and pancreatic juice. Bile is produced in the liver and then stored in the gallbladder. Bile will emulsify the fat. Pancreatic juice contains bicarbonate ions necessary to neutralize acidic chyme, and multiple enzymes for the chemical digestion of carbohydrates, fats, and proteins. The small intestine is nearly 6 meters (21 feet) long and possesses villi that increase its surface area, so chemical digestion is completed and most nutrient absorption occurs before the contents reach the large intestine.

Contents not absorbed enter the *large intestine* through the ileocecal valve. Within the large intestine, mucus is secreted and water and electrolytes are absorbed. Bacteria that inhabit the large intestine can break down some remaining residues, producing some vitamins that are absorbed. Feces composed of water, undigested substances, mucus, and bacteria are formed and stored until elimination.

Procedure A—Oral Cavity and Salivary Glands

1. Study figure 26.1 and the list of structures and descriptions. Examine the head model (sagittal section) and a skull. Locate the following structures:

   oral cavity (mouth)—location of mechanical digestion, some chemical digestion

   vestibule—narrow space between teeth and lips and cheeks

   tongue—muscular organ for food manipulation
   - lingual frenulum (frenulum of tongue)—membranous fold connecting tongue and floor of oral cavity
   - papillae—contain taste buds

   palate—roof of oral cavity
   - hard palate—formed by portions of maxillary and palatine bones
   - soft palate—muscular arch without bone
   - uvula—cone-shaped projection

FIGURE 26.1 The features of the oral cavity.

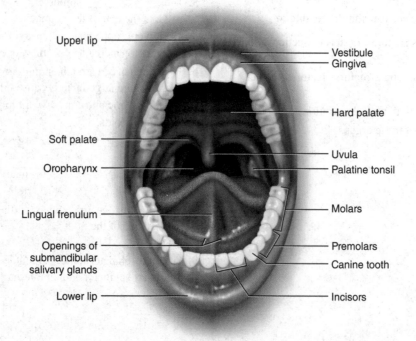

palatine tonsils—lymphatic tissues on lateral walls of pharynx

gingivae (gums)—soft tissue that surrounds neck of tooth and alveolar processes

teeth
- incisors
- canines (cuspids)

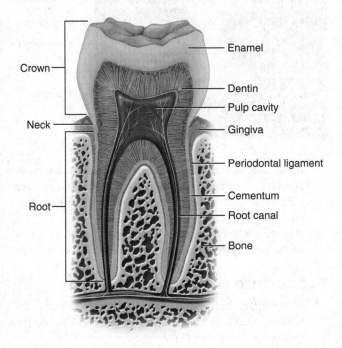

FIGURE 26.2 Longitudinal section of a molar.

Crown
Neck
Root

Enamel
Dentin
Pulp cavity
Gingiva
Periodontal ligament
Cementum
Root canal
Bone

- premolars (bicuspids)
- molars

2. Study figure 26.2 and the list of structures and descriptions. Examine a sectioned tooth and a tooth model. Locate the following features:

   crown—tooth portion projection beyond gingivae
   - enamel—hardest substance of body on surface of crown
   - dentin—living connective tissue forming most of tooth

   neck—junction between crown and root

   root—deep portion of tooth beneath surface
   - pulp cavity—central portion of tooth
   - cementum (cement)—thin surface area of bonelike material
   - root canal—contains blood vessels and nerves

3. Study figures 26.1 and 26.3. Observe the head of the human torso model and locate the following:

   parotid salivary glands—largest glands anterior to ear
   - parotid duct (Stensen's duct)

   submandibular salivary glands—along mandible
   - submandibular duct (Wharton's duct)

   sublingual salivary glands—smallest glands inferior to tongue
   - sublingual ducts (10–12 ducts)

4. Examine a microscopic section of a sublingual gland using low- and high-power magnification. Note the mucous cells that produce mucus and serous cells that produce enzymes. Serous cells sometimes form caps

FIGURE 26.3 The features associated with the salivary glands.

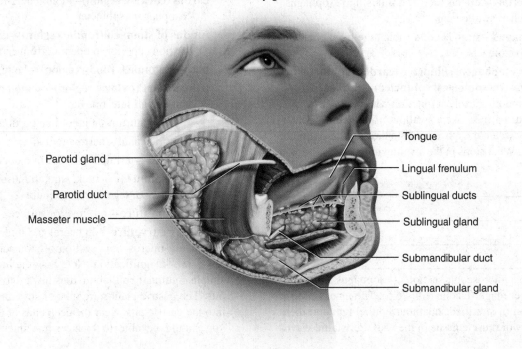

Parotid gland
Parotid duct
Masseter muscle

Tongue
Lingual frenulum
Sublingual ducts
Sublingual gland
Submandibular duct
Submandibular gland

FIGURE 26.4 Micrograph of the sublingual salivary gland (300×).

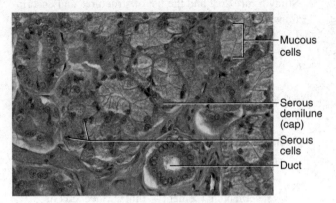

- Mucous cells
- Serous demilune (cap)
- Serous cells
- Duct

FIGURE 26.5 Micrograph of a cross section of the esophagus (10×).

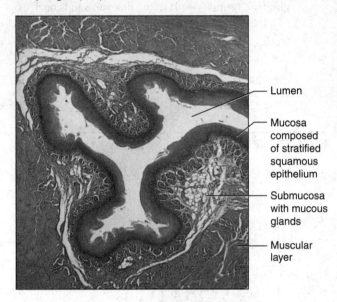

- Lumen
- Mucosa composed of stratified squamous epithelium
- Submucosa with mucous glands
- Muscular layer

(demilunes) around mucous cells. Also note any larger secretory duct surrounded by cuboidal epithelial cells (fig. 26.4).

5. Prepare a labeled sketch of a representative section of a salivary gland in Part A of Laboratory Assessment 26.

Procedure B—Pharynx and Esophagus

1. Examine figure 26.14 on page 246.
2. Observe the human torso model and locate the following features:

   pharynx—connects nasal and oral cavities with larynx

   - nasopharynx—superior to soft palate; without digestive functions
   - oropharynx—posterior to oral cavity
   - laryngopharynx—extends from inferior oropharynx to esophagus

   epiglottis—deflects food and fluids into esophagus when swallowing

   esophagus—straight collapsible tube from pharynx to stomach

   lower esophageal sphincter (cardiac sphincter; gastroesophageal sphincter)—thickened smooth muscle at junction with stomach

3. Have your partner take a swallow from a cup of water. Carefully watch the movements in the anterior region of the neck. What steps in the swallowing process did you observe?

   _____

   _____

   _____

   _____

4. Examine a microscopic section of esophagus wall using low-power magnification (fig. 26.5). The inner lining is composed of stratified squamous epithelium, and there are layers of muscle tissue in the wall. Peristaltic waves

occur from contractions of the muscular layer that move contents through the alimentary canal. Locate some mucous glands in the submucosa. They appear as clusters of lightly stained cells.

5. Prepare a labeled sketch of the esophagus wall in Part A of the laboratory assessment.

Procedure C—Stomach

1. Study figure 26.6 and the list of structures and descriptions. Observe the human torso model and locate the following features of the stomach:

   gastric folds (rugae)—allow stomach to expand to hold food and liquid contents

   cardia (cardiac region)—region near lower esophageal sphincter

   fundus of stomach (fundic region)—dome-shaped region superior to opening from esophagus

   body of stomach (body region)—large main region

   pyloric part (pyloric region)—region near passage into small intestine

   - pyloric antrum—funnel-like portion
   - pyloric canal—narrow region
   - pylorus—terminal passage into duodenum

   pyloric sphincter—thick, smooth muscle valve for chyme passage into duodenum

   lesser curvature—on superior surface

   greater curvature—on lateral and inferior surface

2. Examine a microscopic section of stomach wall using low-power magnification. Note how the inner lining of simple columnar epithelium dips inward to form gastric pits. The gastric glands are tubular structures that open into the gastric pits. Near the deep ends of these glands, you should be able to locate some intensely stained

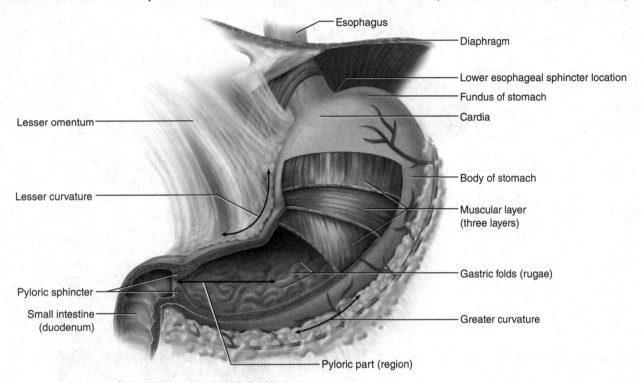

(bluish) chief cells and some lightly stained (pink-ish) parietal cells (fig. 26.7). The parietal cells secrete hydrochloric acid; the chief cells produce pepsinogen, the inactive form of pepsin.

3. Prepare a labeled sketch of a representative section of the stomach wall in Part A of the laboratory assessment.

Procedure D—Pancreas and Liver

1. Study figure 26.8 and the list of structures and descriptions. Observe the human torso model and locate the following structures:

   pancreas—a retroperitoneal organ; secretes an alkaline mixture of digestive enzymes and bicarbonate ions
   - tail of pancreas
   - head of pancreas
   - pancreatic duct
   - accessory pancreatic duct

   liver—has four lobes; produces bile
   - right lobe—largest lobe
   - left lobe—smaller than right lobe
   - quadrate lobe—minor lobe near gallbladder
   - caudate lobe—minor lobe near vena cava

   gallbladder—stores bile

   hepatic ducts—bile ducts in liver

   common hepatic duct—bile duct from liver to cystic duct

FIGURE 26.7 Micrograph of the mucosa of the stomach wall (60×).

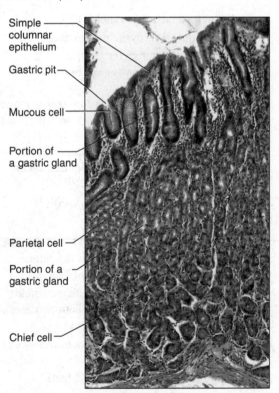

FIGURE 26.8 The features associated with the liver and pancreas.

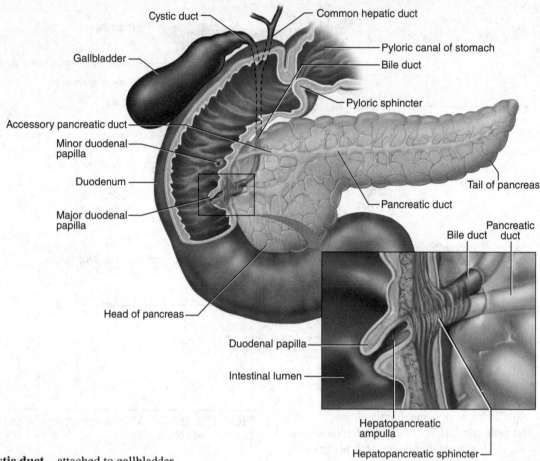

Cystic duct

Common hepatic duct

Gallbladder

Pyloric canal of stomach

Bile duct

Pyloric sphincter

Accessory pancreatic duct

Minor duodenal papilla

Duodenum

Major duodenal papilla

Head of pancreas

Tail of pancreas

Pancreatic duct

Bile duct

Pancreatic duct

Duodenal papilla

Intestinal lumen

Hepatopancreatic ampulla

Hepatopancreatic sphincter

cystic duct—attached to gallbladder

bile duct—duct from cystic duct to duodenum

hepatopancreatic sphincter (sphincter of Oddi)—regulates passage of bile and pancreatic juice

2. Examine the pancreas slide using low-power magnification. Observe the exocrine (acinar) cells that secrete pancreatic juice.

3. Prepare a labeled sketch of a representative section of the pancreas in Part A of the laboratory assessment.

Procedure E—Small and Large Intestines

1. Study figures 26.9, 26.10, and 26.11 and the list of structures and descriptions. Observe the human torso model and locate each of the following features:

small intestine—tube with small diameter and many loops and coils where chemical digestion and absorption of nutrients occur
- duodenum—first 25 cm (10 inches); shortest portion; located retroperitoneal
- jejunum—next nearly 2.5 m (8 feet)
- ileum—last 3.5 m (12 feet); joins with large intestine

mesentery—fold of peritoneal membrane that suspends and supports abdominal viscera

ileocecal valve (sphincter)—regulates contents entering large intestine

large intestine—tube with large diameter nearly 1.5 m (5 feet) long
- large intestinal wall—has 3 unique features
  - teniae coli—3 longitudinal bands of smooth muscle
  - haustra—pouches created by muscle tone of teniae coli
  - epiploic appendages—fatty pouches on outer surfaces
- cecum—blind pouch inferior to ileocecal valve
- appendix (vermiform appendix)—2–7 cm wormlike blind pouch attached to cecum
- ascending colon—course on right side from cecum to transverse colon
- right colic (hepatic) flexure—right-angle turn
- transverse colon—course from ascending to descending colon

242

FIGURE 26.9 The features of the small intestine and associated structures (anterior view). (*Note:* The small intestine is pulled aside to expose the ileocecal junction.)

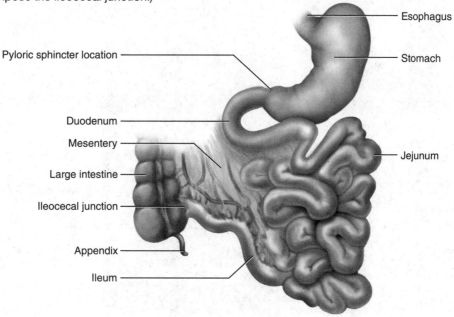

Esophagus

Pyloric sphincter location

Stomach

Duodenum

Mesentery

Jejunum

Large intestine

Ileocecal junction

Appendix

Ileum

FIGURE 26.10 The features of the large intestine (anterior view).

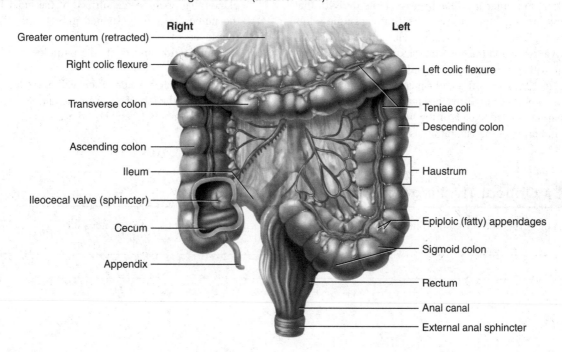

Right

Left

Greater omentum (retracted)

Right colic flexure

Left colic flexure

Transverse colon

Teniae coli

Descending colon

Ascending colon

Ileum

Haustrum

Ileocecal valve (sphincter)

Cecum

Epiploic (fatty) appendages

Appendix

Sigmoid colon

Rectum

Anal canal

External anal sphincter

- left colic (splenic) flexure—right-angle turn
- descending colon—course from transverse to sigmoid colon
- sigmoid colon—S-shaped course to rectum
- rectum—course anterior to sacrum; retains feces until defecation
- anal canal—last few centimeters of large intestine

anal sphincter muscles—encircle anus
- external anal sphincter—the voluntary sphincter of skeletal muscle
- internal anal sphincter—the involuntary sphincter of smooth muscle

anus—external opening

FIGURE 26.11 Normal appendix.

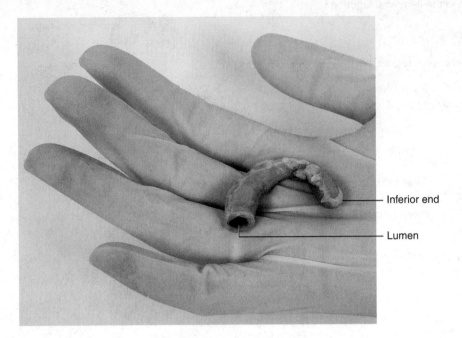

— Inferior end

— Lumen

2. Using low-power magnification, examine a microscopic section of small intestine wall. Identify the mucosa, submucosa, muscular layer, and serosa. Note the villi that extend into the lumen of the tube. Study a single villus using high-power magnification. Note the core of connective tissue and the covering of simple columnar epithelium that contains some lightly stained goblet cells (fig. 26.12). The villi greatly increase the surface area for absorption of digestive products.

3. Prepare a labeled sketch of the wall of the small intestine in Part A of the laboratory assessment.

4. Examine a microscopic section of large intestine wall. Note the lack of villi. Also note the tubular mucous glands that open on the surface of the inner lining and the numerous lightly stained goblet cells. Locate the four layers of the wall (fig. 26.13). The mucus functions as a lubrication and holds the particles of fecal matter together.

5. Prepare a labeled sketch of the wall of the large intestine in Part A of the laboratory assessment.

6. Complete Parts B, C, and D of the laboratory assessment.

Critical Thinking Activity

How is the structure of the small intestine better adapted for absorption than that of the large intestine?

_____

_____

FIGURE 26.12 Micrograph of the small intestine wall (40×).

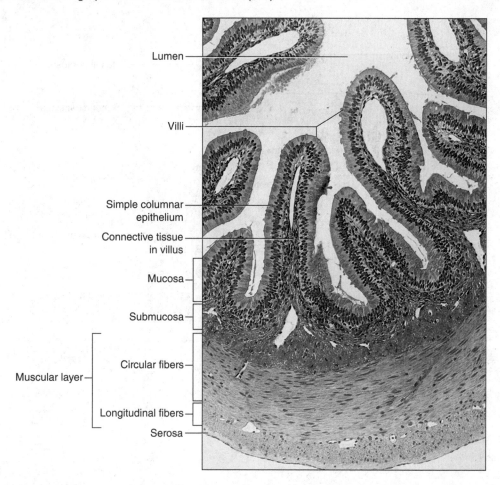

Lumen

Villi

Simple columnar epithelium

Connective tissue in villus

Mucosa

Submucosa

Muscular layer

Circular fibers

Longitudinal fibers

Serosa

FIGURE 26.13 Micrograph of the large intestine wall (64×).

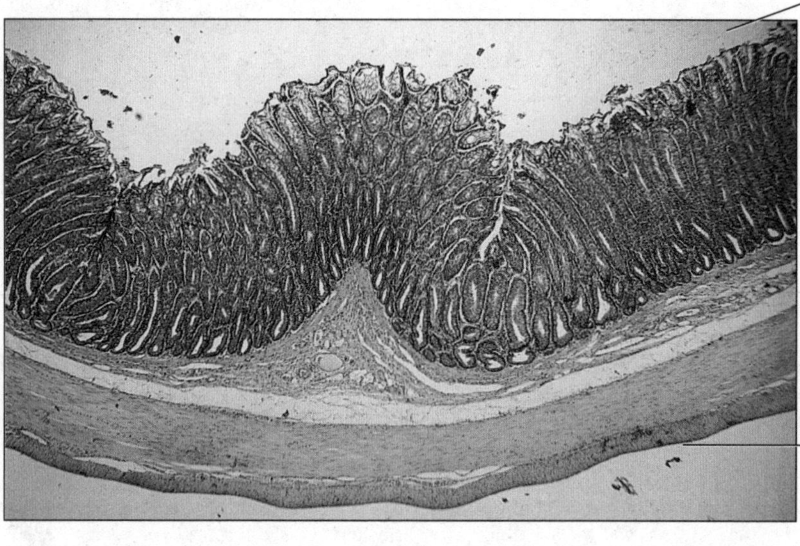

Lumen

Mucosa with mucous glands

Submucosa

Muscular layer (circular and longitudinal)

Serosa

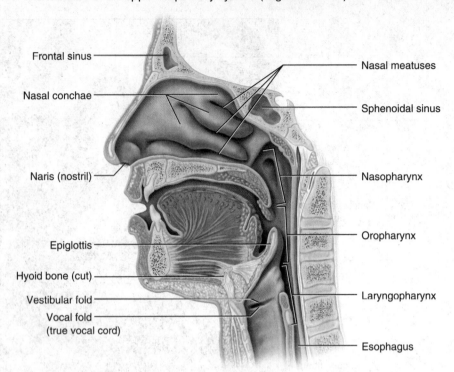

Frontal sinus

Nasal conchae

Naris (nostril)

Epiglottis

Hyoid bone (cut)

Vestibular fold

Vocal fold
(true vocal cord)

Nasal meatuses

Sphenoidal sinus

Nasopharynx

Oropharynx

Laryngopharynx

Esophagus

Name _____

Date _____

Section _____

The ⚠ corresponds to the indicated outcome(s) found at the beginning of the laboratory exercise.

Digestive Organs

Part A   Assessments

In the space that follows, sketch a representative area of the organ indicated. Label any of the structures observed and indicate the magnification used for each sketch. ⚠

Salivary gland (_____×)	Esophagus (_____×)
Stomach wall (_____×)	Pancreas (_____×)
Small intestine wall (_____×)	Large intestine wall (_____×)

Part B Assessments

Identify the numbered features in figures 26.15, 26.16, 26.17, and 26.18.

FIGURE 26.15   Label the features of the stomach and nearby regions in this frontal section of a cadaver (anterior view). 🄰

Anatomy & Physiology | REVEALED
aprevealed.com

Pancreas

Duodenal papilla

5
6
7
8
9
Gastric folds (rugae)
10

1
2
3
4

FIGURE 26.16   Label the features associated with the liver and pancreas. 🄰

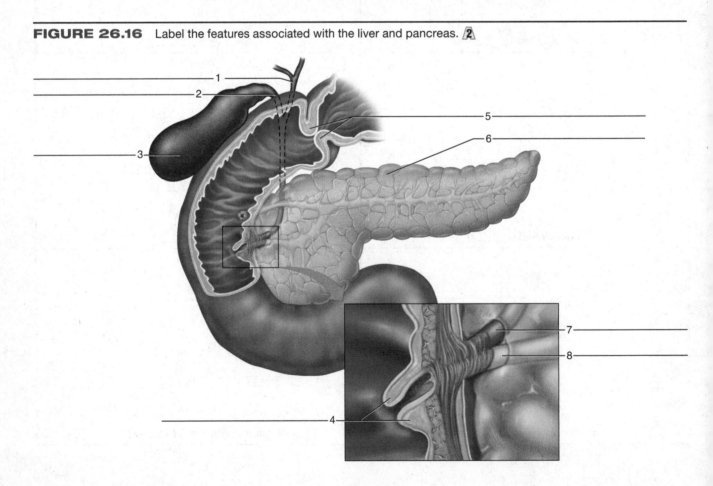

1
2
3
5
6
7
8
4

248

FIGURE 26.17 Label the digestive structures of this abdominopelvic cavity of a cadaver (anterior view). 🄰

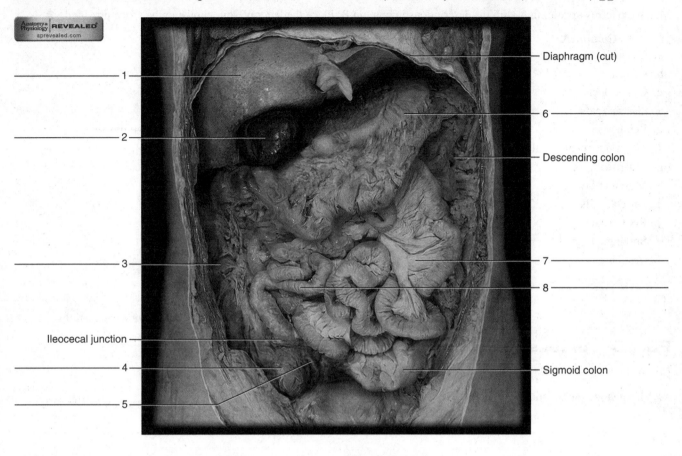

1

2

3

Ileocecal junction

4

5

Diaphragm (cut)

6

Descending colon

7

8

Sigmoid colon

FIGURE 26.18 Label the features of the large intestine. 🄰

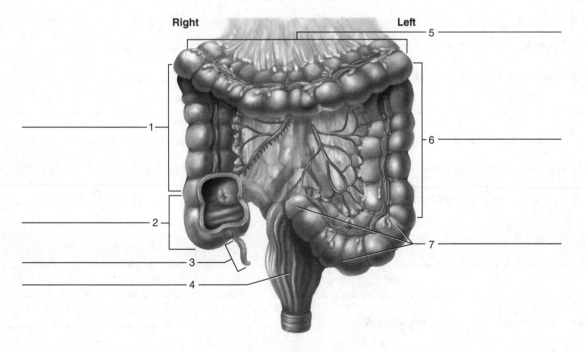

Right

Left

1

2

3

4

5

6

7

249

Part C Assessments

Match the terms in column A with the descriptions in column B. Place the letter of your choice in the space provided. ⚠3

Column A	Column B
a. Cardia	_____ **1.** Smallest of major salivary glands
b. Crown	_____ **2.** Secrete hydrochloric acid into stomach
c. Cystic duct	_____ **3.** Last section of small intestine
d. Gastric folds	_____ **4.** Region of stomach near lower esophageal sphincter
e. Ileum	_____ **5.** Contains blood vessels and nerves in a tooth
f. Major duodenal papilla	_____ **6.** Responsible for peristaltic waves
g. Mucosa	_____ **7.** Allow stomach to expand
h. Muscular layer	_____ **8.** Space between teeth, cheeks, and lips
i. Parietal cells	_____ **9.** Attached to gallbladder
j. Root canal	_____ **10.** Portion of tooth projecting beyond gingivae
k. Sublingual gland	_____ **11.** Layer nearest lumen of alimentary canal
l. Vestibule	_____ **12.** Common opening region for bile and pancreatic secretions

Part D Assessments

Complete the following:

1. Summarize the functions of the oral cavity. ⚠A _____

_____

_____

2. Summarize the functions of the esophagus. ⚠A _____

_____

3. Name the valve that prevents regurgitation of food from the duodenum back into the stomach. ⚠2 _____

4. Summarize the functions of the stomach. ⚠A _____

_____

_____

5. Name the valve that controls the movement of material between the small and large intestines. ⚠2 _____

6. Summarize the functions of the small intestine. ⚠A _____

_____

_____

_____

7. Summarize the functions of the large intestine. ⚠A _____

_____

_____

Action of a Digestive Enzyme

Purpose of the Exercise

To investigate the action of amylase and the effect of heat on its enzymatic activity.

Materials Needed

0.5% amylase solution*
Beakers (50 and 500 mL)
Distilled water
Funnel
Pipettes (1 and 10 mL)
Pipette rubber bulbs
0.5% starch solution
Graduated cylinder (10 mL)
Test tubes
Test-tube clamps
Wax marker
Iodine-potassium-iodide solution

Medicine dropper
Ice
Water bath, 37°C (98.6°F)
Porcelain test plate
Benedict's solution
Hot plates
Test-tube rack
Thermometer

For Alternative Procedure Activity:
Small disposable cups
Distilled water

*The amylase should be free of sugar for best results; a low-maltose solution of amylase yields good results. See the Appendix.

Safety

▶ Use only a mechanical pipetting device (never your mouth). Use pipettes with rubber bulbs or dropping pipettes.
▶ Wear safety glasses when working with acids and when heating test tubes.
▶ Use test-tube clamps when handling hot test tubes.
▶ If an open flame is used for heating the test solutions, keep clothes and hair away from the flame.
▶ Review all the safety guidelines inside the front cover.
▶ If student saliva is used as a source of amylase, it is important that students wear disposable gloves and only handle their own materials.
▶ Use an appropriate disinfectant to wash the laboratory tables before and after the procedures.
▶ Dispose of chemicals according to appropriate directions.
▶ Wash your hands before leaving the laboratory.

Learning Outcomes

After completing this exercise, you should be able to

1. Test a solution for the presence of starch or the presence of sugar.
2. Explain the action of amylase.
3. Test the effects of varying temperatures on the activity of amylase.

Pre-Lab

Carefully read the introductory material and examine the entire lab. Be familiar with enzyme structure and activity from lecture or the textbook. Answer the pre-lab questions.

Pre-Lab Questions: Select the correct answer for each of the following questions:

1. The enzyme amylase accelerates the reaction of changing
   a. polysaccharides into monosaccharides.
   b. disaccharides into monosaccharides.
   c. starch into disaccharides.
   d. disaccharides into glucose.

2. Amylase originates from
   a. salivary glands and stomach.
   b. salivary glands and pancreas.
   c. esophagus and stomach.
   d. pancreas and intestinal mucosa.

3. The optimum pH for amylase activity is
   a. 0–14.   b. 2.
   c. 6.8–7.0.   d. 7–8.

4. After an enzyme reaction is completed, the enzyme
   a. is still available.
   b. was structurally changed.
   c. was denatured.
   d. is part of the enzyme-substrate complex.

5. As temperatures decrease, enzyme activity
   a. increases.   b. remains unchanged.
   c. ceases.   d. decreases.

The laboratory exercise is designed to examine the relationship between enzymes and chemical digestion. *Enzymes* are proteins that serve as *biological catalysts*. Acting as a catalyst, an enzyme will modify and rapidly increase the rate of a particular chemical reaction without being consumed in the reaction, which enables the enzyme to function repeatedly. A particular enzyme will catalyze a specific reaction. Various metabolic pathways and chemical digestive processes are accelerated by numerous enzymatic reactions. The **lock-and-key model** of enzyme activity involves an *enzyme* and *substrate* (substance acted upon by the enzyme) temporarily combining as an *enzyme-substrate complex* and ending with a reaction *product* and the original *enzyme* (fig. 27.1).

Enzymes are globular proteins with a three-dimensional structure determined by various chemical bonds including the hydrogen bond. For the enzyme to function as the catalyst, the shape is critical at the active site of the enzyme for the enzyme-substrate complex to take place properly. If the environmental temperature, including the body temperature, rises to a certain level, the three-dimensional shape of the enzyme is modified as hydrogen bonds start breaking and the enzyme becomes *denatured*. Although the primary structure of the amino acid sequence remains unaltered, the secondary and tertiary structures have been altered. A denatured protein could regain its shape unless the three-dimensional shape is destroyed when conditions become too extreme, and the enzyme becomes irreversibly denatured.

The digestive enzyme in salivary secretions is called *salivary amylase*. Pancreatic amylase is secreted among several other pancreatic enzymes. A bacterial extraction of amylase is available for laboratory experiments. This enzyme catalyzes the reaction of changing starch molecules into sugar (disaccharide) molecules, which is the first step in the digestion of complex carbohydrates.

As in the case of other enzymes, amylase is a protein catalyst. Its activity is affected by exposure to certain environmental factors, including various temperatures, pH, radiation, and electricity. As temperatures increase, faster chemical reactions occur as the collisions of molecules happen at a greater frequency. Eventually, temperatures increase to a point that the enzyme is denatured, and the rate of the enzyme activity rapidly declines. As temperatures decrease, enzyme activity also decreases due to fewer collisions of the molecules; however, the colder temperatures do not denature the enzyme. Normal body temperature provides an environment for enzyme activity near the optimum for enzymatic reactions.

The pH of the environment where enzymes are secreted also has a major influence on the enzyme reactions. The optimum pH for amylase activity is between 6.8 and 7.0, typical of salivary secretions. When salivary amylase arrives in the stomach, hydrochloric acid deactivates the enzyme, diminishing any further chemical digestion of remaining starch in the stomach. However, pancreatic amylase is secreted into the small intestine, where an optimum pH for amylase is once more provided. Other enzymes in the digestive system have different optimum pH ranges of activity compared to amylase. For example, pepsin from stomach secretions has an optimum activity around pH 2, while trypsin from pancreatic secretions operates best around pH 7–8.

Procedure A—Amylase Activity

1. Study the lock-and-key model example specific for amylase action on starch digestion (fig. 27.2).
2. Examine the locations where starch is digested into glucose (fig. 27.3).
3. Mark three clean test tubes as *tubes 1, 2,* and *3,* and prepare the tubes as follows (fig. 27.4):

   Tube 1: **Add 6 mL of amylase solution.**

   Tube 2: **Add 6 mL of starch solution.**

   Tube 3: **Add 5 mL of starch solution and 1 mL of amylase solution.**

FIGURE 27.1 Lock-and-key model of an enzyme-catalyzed reaction ($E + S \rightarrow E\text{--}S \rightarrow P + E$). (Many enzyme-catalyzed reactions, as depicted here, are reversible.) In the forward reaction (dark-shaded arrows), (a) the shapes of the substrate molecules fit the shape of the enzyme's active site. (b) When the substrate molecules temporarily combine with the enzyme, a chemical reaction occurs. (c) The result is a product molecule and an unaltered enzyme. The active site changes shape somewhat as the substrate binds, such that formation of the enzyme-substrate complex is more like a hand fitting into a glove, which has some flexibility, than a key fitting into a lock.

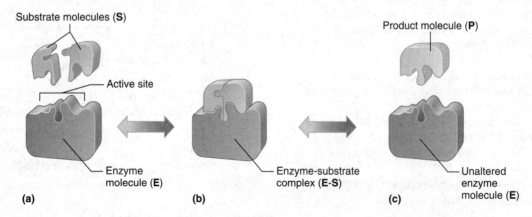

Substrate molecules (**S**)

Active site

Enzyme molecule (**E**)

(a)

Enzyme-substrate complex (**E-S**)

(b)

Product molecule (**P**)

Unaltered enzyme molecule (**E**)

(c)

FIGURE 27.2 Lock-and-key model of enzyme action of amylase on starch digestion.

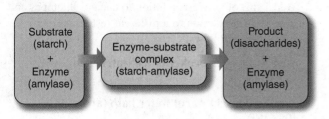

Alternative Procedure Activity

Human saliva could be used as a source of amylase solutions instead of bacterial amylase preparations. Collect about 5 mL of saliva into a small disposable cup. Add an equal amount of distilled water and mix them together for the amylase solutions during the laboratory procedures. Be sure to follow all of the safety guidelines.

FIGURE 27.3 Flowchart of starch digestion.

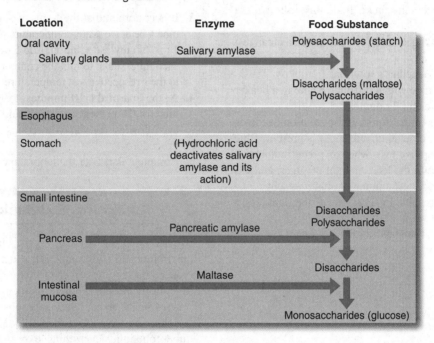

FIGURE 27.4 Test tubes prepared for testing amylase activity.

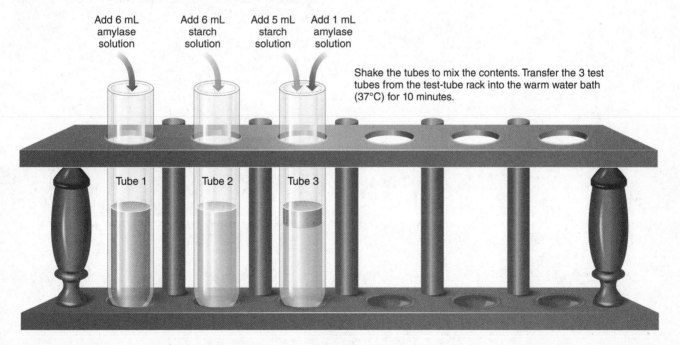

Add 6 mL amylase solution

Add 6 mL starch solution

Add 5 mL starch solution

Add 1 mL amylase solution

Shake the tubes to mix the contents. Transfer the 3 test tubes from the test-tube rack into the warm water bath (37°C) for 10 minutes.

Tube 1   Tube 2   Tube 3

4. Shake the tubes well to mix the contents and place them in a warm water bath, 37°C (98.6°F), for 10 minutes.
5. At the end of the 10 minutes, test the contents of each tube for the presence of starch. To do this, follow these steps:
   a. Place 1 mL of the solution to be tested in a depression of a porcelain test plate.
   b. Next add one drop of iodine-potassium-iodide solution and note the color of the mixture. If the solution becomes blue-black, starch is present.
   c. Record the results in Part A of Laboratory Assessment 27.
6. Test the contents of each tube for the presence of sugar (disaccharides in this instance). To do this, follow these steps:
   a. Place 1 mL of the solution to be tested in a clean test tube.
   b. Add 1 mL of Benedict's solution.
   c. Place the test tube with a test-tube clamp in a beaker of boiling water for 2 minutes.
   d. Note the color of the liquid. If the solution becomes green, yellow, orange, or red, sugar is present. Blue indicates a negative test, whereas green indicates a positive test with the least amount of sugar, and red indicates the greatest amount of sugar present.
   e. Record the results in Part A of the laboratory assessment.
7. Complete Part A of the laboratory assessment.

Procedure B—Effect of Heat

1. Mark three clean test tubes as *tubes 4, 5,* and *6.*
2. Add 1 mL of amylase solution to each of the tubes and expose each solution to a different test temperature for 3 minutes as follows:

   Tube 4: **Place in beaker of ice water (about 0°C [32°F]).**

   Tube 5: **Place in warm water bath (about 37°C [98.6°F]).**

   Tube 6: **Place in beaker of boiling water (about 100°C [212°F]). Use a test-tube clamp.**

3. It is important that the 5 mL of starch solution added to tube 4 be at ice-water temperature before it is added to the 1 mL of amylase solution. Add 5 mL of starch solution to each tube, shake to mix the contents, and return the tubes to their respective test temperatures for 10 minutes.
4. At the end of the 10 minutes, test the contents of each tube for the presence of starch and the presence of sugar by following the same directions as in steps 5 and 6 in Procedure A.
5. Complete Part B of the laboratory assessment.

Learning Extension Activity

Devise an experiment to test the effect of some other environmental factor on amylase activity. For example, you might test the effect of a strong acid by adding a few drops of concentrated hydrochloric acid to a mixture of starch and amylase solutions. Be sure to include a control in your experimental plan. That is, include a tube containing everything except the factor you are testing. Then you will have something with which to compare your results. *Carry out your experiment only if it has been approved by the laboratory instructor.*

Name _____

Date _____

Section _____

The Ⓐ corresponds to the indicated outcome(s) found at the beginning of the laboratory exercise.

Action of a Digestive Enzyme

Part A   Amylase Activity Assessments

1. Test results: ▲

Tube	Starch	Sugar
1 Amylase solution		
2 Starch solution		
3 Starch-amylase solution		

2. Complete the following:

   a. Explain the reason for including tube 1 in this experiment. ▲ _____

_____

_____

_____

   b. What is the importance of tube 2? ▲ _____

_____

_____

_____

   c. What do you conclude from the results of this experiment? ▲ _____

_____

_____

_____

Part B   Effect of Heat Assessments

1. Test results: 🔺3

Tube	Starch	Sugar
4 0°C (32°F)		
5 37°C (98.6°F)		
6 100°C (212°F)		

2. Complete the following:

   a. What do you conclude from the results of this experiment? 🔺3 _____

   _____

   _____

   _____

   _____

   _____

   b. If digestion failed to occur in one of the tubes in this experiment, how can you tell if the amylase was destroyed by

   the factor being tested or if the amylase activity was simply inhibited by the test treatment? 🔺3

   _____

   _____

   _____

   _____

   _____

Critical Thinking Assessment

What test result would occur if the amylase you used contained sugar? _____ Would your results
be as valid? Explain your answer. 🔺2

_____

_____

_____

_____

Urinary Organs

Purpose of the Exercise

To review the structure of urinary organs, to dissect a kidney, and to observe the major structures of a nephron.

Materials Needed

Human torso model
Kidney model
Preserved pig (or sheep) kidney
Dissecting tray
Dissecting instruments
Long knife
Compound light microscope
Prepared microscope slide of the following:
    Kidney section
    Ureter, cross section
    Urinary bladder
    Urethra, cross section

⚠ Safety

▶ Wear disposable gloves when working on the kidney dissection.
▶ Dispose of the kidney and gloves as directed by your laboratory instructor.
▶ Wash the dissecting tray and instruments as instructed.
▶ Wash your laboratory table.
▶ Wash your hands before leaving the laboratory.

Learning Outcomes

After completing this exercise, you should be able to

1. Locate and identify the major structures of a kidney.
2. Identify and sketch the major structures of a nephron.
3. Trace the path of filtrate through a renal nephron.
4. Trace the path of blood through the renal blood vessels.
5. Identify and sketch the structures of a ureter and a urinary bladder wall.
6. Trace the path of urine flow through the urinary system.

Pre-Lab

Carefully read the introductory material and examine the entire lab. Be familiar with the structures and functions of the urinary organs from lecture or the textbook. Answer the pre-lab questions.

Pre-Lab Questions: Select the correct answer for each of the following questions:

1. When comparing the position of the two kidneys,
   a. they are at the same level.
   b. the right kidney is slightly superior to the left kidney.
   c. the right kidney is slightly inferior to the left kidney.
   d. the right kidney is anterior to the left kidney.

2. Cortical nephrons represent about _____ of the nephrons.
   a. 0 %   b. 100 %
   c. 20 %   d. 80 %

3. Which of the following does *not* represent one of the processes in urine formation?
   a. secretion of renin
   b. glomerular filtration
   c. tubular reabsorption
   d. tubular secretion

4. The _____ arteries and veins are located in the corticomedullary junction.
   a. arcuate   b. renal
   c. cortical radiate   d. peritubular

5. The _____ is the tube from the kidney to the urinary bladder.
   a. urethra   b. ureter
   c. renal pelvis   d. renal column

6. The trigone is a triangular, funnel-like region of the
   a. renal cortex.   b. renal medulla.
   c. urethra.   d. urinary bladder.

7. The external urethral sphincter is composed of involuntary smooth muscle.
   True _____   False _____

8. Contractions of the detrusor muscle provide the force during micturition (urination).
   True _____   False _____

The two kidneys are the primary organs of the urinary system. They are located in the abdominal cavity, against the posterior wall and behind the parietal peritoneum (retroperitoneal). Masses of adipose tissue associated with the kidneys hold them in place at a vertebral level between T12 and L3. The right kidney is slightly inferior due to the large mass of the liver near its superior border. Ureters force urine by means of peristaltic waves into the urinary bladder, which temporarily stores urine. The urethra conveys urine to the outside of the body.

Each kidney contains over 1 million nephrons, which serve as the basic structural and functional units of the kidney. A glomerular capsule, proximal convoluted tubule, nephron loop, and distal convoluted tubule compose the microscopic, multicellular structure of a relatively long nephron tubule, which drains into a collecting duct. Approximately 80% of the nephrons are cortical nephrons with short nephron loops, while the remaining represent juxtamedullary nephrons, with long nephron loops extending deeper into the renal medulla. An elaborate network of blood vessels surrounds the entire nephron. Glomerular filtration, tubular reabsorption, and tubular secretion represent three processes resulting in urine as the final product.

A variety of functions occur in the kidneys. They remove metabolic wastes from the blood; help regulate blood volume, blood pressure, and pH of blood; control water and electrolyte concentrations; and secrete renin and erythropoietin. Renin functions in regulation of blood pressure and erythropoietin helps regulate red blood cell production.

Procedure A—Kidney Structure

1. Study figures 28.1 and 28.2 and the list of structures and descriptions.
2. Observe the human torso model and the kidney model. Locate the following:

    kidneys—paired retroperitoneal organs

    ureters—paired tubular organs about 25 cm long that transport urine from kidney to urinary bladder

    urinary bladder—single muscular storage organ for urine

    urethra—conveys urine from urinary bladder to external urethral orifice

    renal sinus—hollow chamber of kidney

    renal pelvis—funnel-shaped sac at superior end of ureter; receives urine from major calyces

    - major calyces—2–3 major converging branches into renal pelvis; receive urine from minor calyces
    - minor calyces—small subdivisions of major calyces; receive urine from papillary ducts within renal papillae

    renal medulla—deep region of kidney

    - renal pyramids—6–10 conical regions composing most of renal medulla

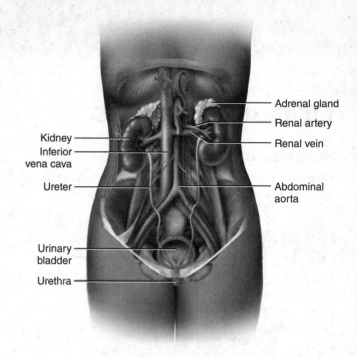

FIGURE 28.1 The urinary system and associated structures.

- renal papillae—projections at ends of renal pyramids with openings into minor calyces

renal cortex—superficial region of kidney

renal columns—extensions of renal cortical tissue between renal pyramids

nephrons—functional units of kidneys

- cortical nephrons—80% of nephrons close to surface
- juxtamedullary nephrons—20% of nephrons close to medulla

3. To observe the structure of a kidney, follow these steps:

    a. Obtain a preserved pig or sheep kidney and rinse it with water to remove as much of the preserving fluid as possible.

    b. Carefully remove any adipose tissue from the surface of the specimen.

    c. Locate the following features:

    fibrous (renal) capsule—protective membrane that encloses kidney

    hilum of kidney—indented region containing renal artery and vein and ureter

    renal artery—large artery arising from abdominal aorta

    renal vein—large vein that drains into inferior vena cava

    ureter—transports urine from renal pelvis to urinary bladder

    d. Use a long knife to cut the kidney in half longitudinally along the frontal plane, beginning on the convex border.

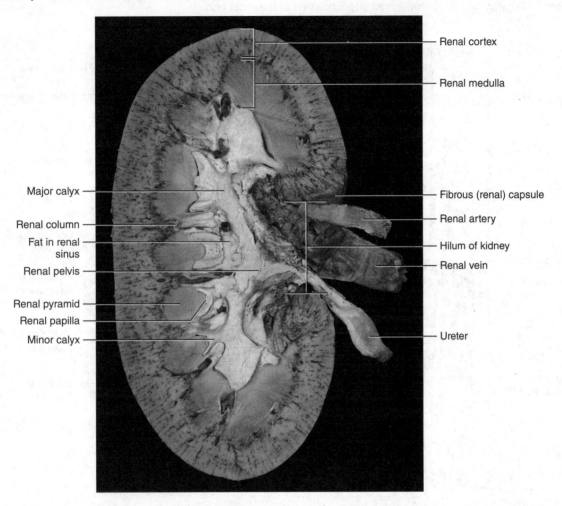

Labels on figure:
- Renal cortex
- Renal medulla
- Fibrous (renal) capsule
- Renal artery
- Hilum of kidney
- Renal vein
- Ureter
- Major calyx
- Renal column
- Fat in renal sinus
- Renal pelvis
- Renal pyramid
- Renal papilla
- Minor calyx

e. Rinse the interior of the kidney with water, and using figure 28.2 as a reference, locate the following:

renal pelvis

- major calyces
- minor calyces

renal cortex

renal columns

renal medulla

- renal pyramids
- renal papillae

4. Complete Part A of Laboratory Assessment 28.

Procedure B—Renal Blood Vessels and Nephrons

1. Study figures 28.3 and 28.4 illustrating renal blood vessels and structures of a nephron. The arrows within the figures indicate blood flow and tubular fluid flow.

2. Obtain a microscope slide of a kidney section and examine it using low-power magnification. Locate the *renal capsule,* the *renal cortex* (which appears some-

what granular and may be more darkly stained than the other renal tissues), and the *renal medulla* (fig. 28.5).

3. Examine the renal cortex using high-power magnification. Locate a *renal corpuscle.* These structures appear as isolated circular areas. Identify the *glomerulus,* the capillary cluster inside the corpuscle, and the *glomerular (Bowman's) capsule,* which appears as a clear area surrounding the glomerulus. A glomerulus and a glomerular capsule compose a *renal corpuscle.* Also note the numerous sections of renal tubules that occupy the spaces between renal corpuscles (fig. 28.5a).

4. Prepare a labeled sketch of a representative section of the renal cortex in Part B of the laboratory assessment.

5. Examine the renal medulla using high-power magnification. Identify longitudinal and cross-sectional views of various collecting ducts. These ducts are lined with simple epithelial cells, which vary in shape from squamous to cuboidal (fig. 28.5b).

6. Prepare a labeled sketch of a representative section of the renal medulla in Part B of the laboratory assessment.

7. Complete Part C of the laboratory assessment.

FIGURE 28.3 Renal blood vessels associated with cortical and juxtamedullary nephrons. The arrows indicate the flow of blood. Blood from the renal artery flows through a segmental artery and an interlobar artery to the arcuate artery. Blood from an arcuate vein flows through an interlobar vein to the renal vein.

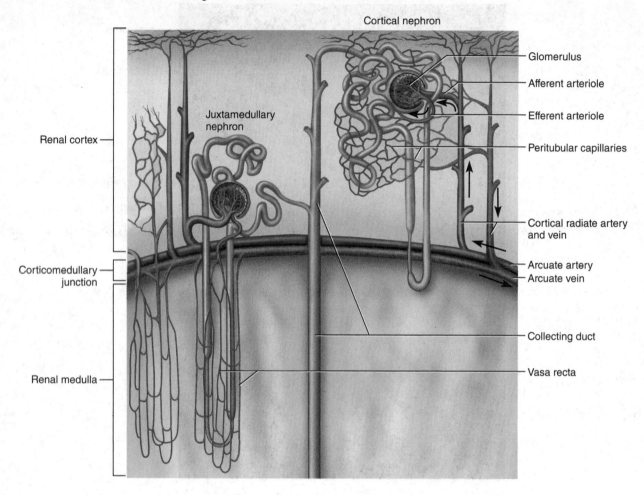

Cortical nephron

Glomerulus

Afferent arteriole

Efferent arteriole

Peritubular capillaries

Juxtamedullary nephron

Renal cortex

Cortical radiate artery and vein

Corticomedullary junction

Arcuate artery
Arcuate vein

Collecting duct

Renal medulla

Vasa recta

FIGURE 28.4 Structure of a nephron with arrows indicating the flow of tubular fluid. The corticomedullary junction is typical for a cortical nephron.

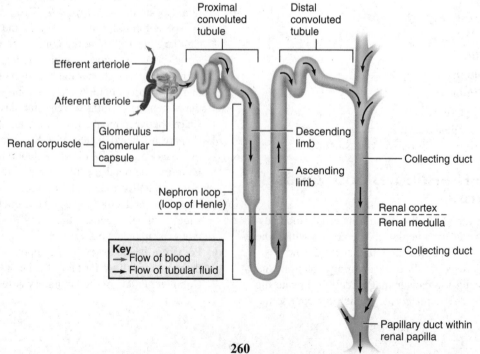

Proximal convoluted tubule

Distal convoluted tubule

Efferent arteriole

Afferent arteriole

Renal corpuscle
Glomerulus
Glomerular capsule

Descending limb

Collecting duct

Ascending limb

Nephron loop (loop of Henle)

Renal cortex
Renal medulla

Collecting duct

Key
→ Flow of blood
→ Flow of tubular fluid

Papillary duct within renal papilla

FIGURE 28.5   (a) Micrograph of a section of the renal cortex (220×). (b) Micrograph of a section of the renal medulla (80×).

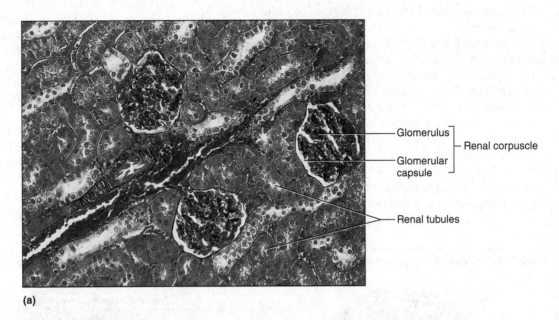

Glomerulus ⎤
⎬ Renal corpuscle
Glomerular capsule ⎦

Renal tubules

(a)

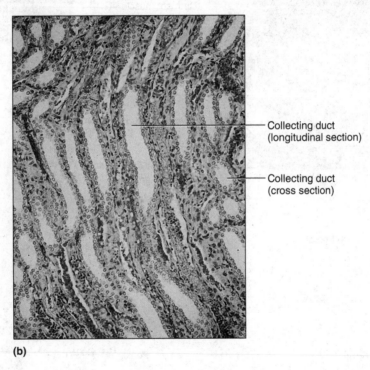

Collecting duct (longitudinal section)

Collecting duct (cross section)

(b)

Procedure C—Ureter, Urinary Bladder, and Urethra

The *ureters* are paired tubes about 25 cm (10 inches) long extending from the renal pelvis to the *ureteral openings (ureteric orifices)* of the *urinary bladder*. Peristaltic waves, created from rhythmic contractions of smooth muscle, occur about every 30 seconds, transporting the urine toward the urinary bladder. The bladder possesses mucosal folds called *rugae* that enable it to distend for temporary urine storage. Coarse bundle layers of smooth muscle compose the somewhat broad *detrusor muscle*. The triangular floor of the bladder, the *trigone,* is bordered by the two ureteral openings and the opening into the urethra, which is surrounded by the *internal urethral sphincter.* The internal urethral sphincter is formed by a thickened region of the detrusor muscle (fig. 28.6).

Although the bladder can hold about 600 mL of urine, the desire to urinate from the pressure of a stretched bladder occurs when the bladder has about 150–200 mL of urine. During *micturition (urination)* the powerful detrusor muscle contracts, and the internal urethral sphincter is forced open. The *external urethral sphincter,* at the level of the urogenital diaphragm, is under somatic control. Because the external urethral sphincter is skeletal muscle, there is considerable voluntary control over the voiding of urine.

The *urethra* extends from the exit of the urinary bladder to the *external urethral orifice.* In the female the urethra is about 4 cm (1.5 inches) long. In the male, the urethra includes passageways for both urine and semen and extends the length of the penis. The male urethra can be divided into three sections: prostatic urethra, membranous (intermediate) urethra, and spongy urethra (fig. 28.6*b*). A total length of

FIGURE 28.6 Ureters and frontal sections of urinary bladder and urethra of (a) female and (b) male.

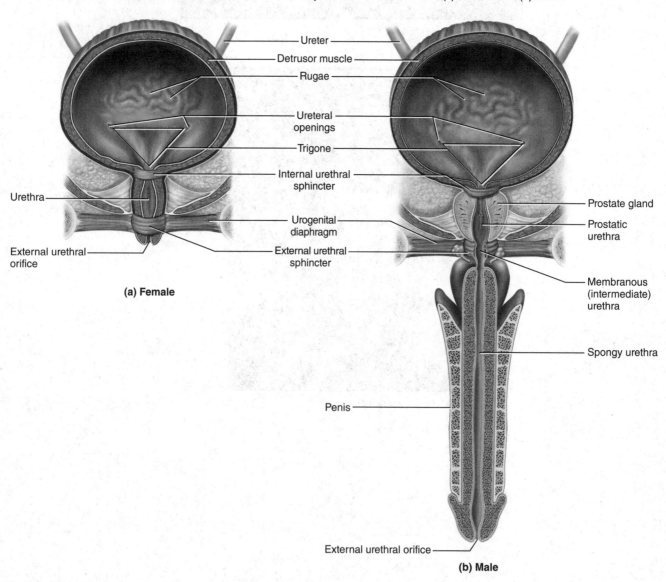

(a) Female

(b) Male

about 20 cm (8 inches) would be characteristic for the urethra of a male.

1. Obtain a microscope slide of a cross section of a ureter and examine it using low-power magnification. Locate the *mucous coat* layer next to the lumen. Examine the middle *muscular coat* composed of longitudinal and circular smooth muscle cells responsible for the peristaltic waves that propel urine from the kidneys to the urinary bladder. The outer *fibrous coat,* composed of connective tissue, secures the ureter in the retroperitoneal position (fig. 28.7).

2. Examine the mucous coat using high-power magnification. The specialized tissue is transitional epithelium, which allows changes in its thickness when unstretched and stretched.

3. Prepare a labeled sketch of a ureter in Part D of the laboratory assessment.

4. Obtain a microscope slide of a segment of the wall of a urinary bladder and examine it using low-power magnification. Examine the *mucous coat* next to the lumen and the *submucous coat* composed of connective tissue just beneath the mucous coat. Examine the *muscular coat* composed of bundles of smooth muscle fibers interlaced in many directions. This thick muscular layer is called the *detrusor muscle* and functions in the elimination of urine. Also note the outer *serous coat* of connective tissue (fig. 28.8).

5. Examine the mucous coat using high-power magnification. The tissue is transitional epithelium, which allows changes in its thickness from unstretched when the bladder is empty to stretched when the bladder distends with urine.

FIGURE 28.7 Micrograph of a cross section of a ureter (75×).

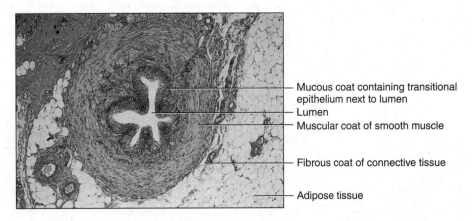

- Mucous coat containing transitional epithelium next to lumen
- Lumen
- Muscular coat of smooth muscle
- Fibrous coat of connective tissue
- Adipose tissue

FIGURE 28.8 Micrograph of a segment of the human urinary bladder wall (6×).

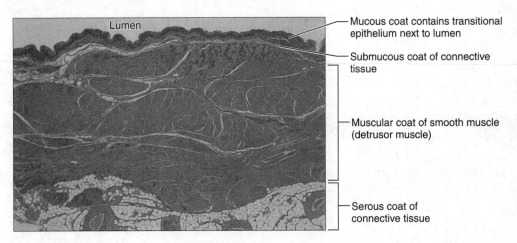

Lumen

- Mucous coat contains transitional epithelium next to lumen
- Submucous coat of connective tissue
- Muscular coat of smooth muscle (detrusor muscle)
- Serous coat of connective tissue

6. Prepare a labeled sketch of a segment of the urinary bladder wall in Part D of the laboratory assessment.

7. Obtain a microscope slide of a cross section of a urethra and examine it using low-power magnification. Locate the muscular coat composed of smooth muscle fibers. Locate groups of mucous glands, called *urethral glands,* in the urethral wall that secrete protective mucus into the lumen of the urethra. Examine the mucous coat using high-power magnification. The mucous membrane is composed of a type of stratified epithelium. The specific type of stratified epithelium varies from transitional, to stratified columnar, to stratified squamous epithelium between the urinary bladder and the external urethral orifice. Depending upon where the section of the urethra was taken for the preparation of the microscope slide, the type of stratified epithelial tissue represented could vary (fig. 28.9).

8. Complete Part E of the laboratory assessment.

FIGURE 28.9 Cross section through the urethra (10×).

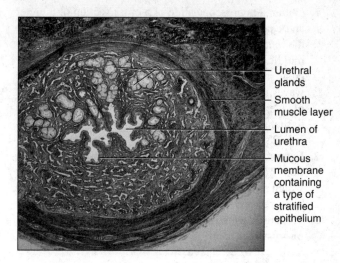

Urethral glands

Smooth muscle layer

Lumen of urethra

Mucous membrane containing a type of stratified epithelium

Name _____

Date _____

Section _____

The Ⓐ corresponds to the indicated outcome(s) found at the beginning of the laboratory exercise.

Urinary Organs

Part A Assessments

1. Label the features indicated in figure 28.10 of a kidney (frontal section).

FIGURE 28.10   Label the structures in the frontal section of a kidney. Ⓐ

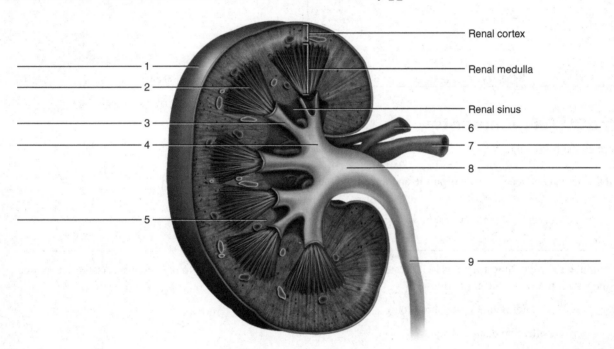

2. Match the terms in column A with the descriptions in column B. Place the letter of your choice in the space provided. Ⓐ

Column A	Column B
a. Calyces	_____ **1.** Superficial region around the renal medulla
b. Hilum of kidney	_____ **2.** Branches of renal pelvis to renal papillae
c. Nephron	_____ **3.** Conical mass of tissue within renal medulla
d. Renal column	_____ **4.** Projection with tiny openings into a minor calyx
e. Renal cortex	_____ **5.** Hollow chamber within kidney
f. Renal papilla	_____ **6.** Microscopic functional unit of kidney
g. Renal pelvis	_____ **7.** Located between renal pyramids
h. Renal pyramid	_____ **8.** Superior funnel-shaped end of ureter inside the renal sinus
i. Renal sinus	_____ **9.** Medial depression for blood vessels and ureter to enter kidney chamber

Part B Assessments

Sketch a representative section of the renal cortex and the renal medulla. Label the glomerulus, glomerular capsule, and sections of renal tubules in the renal cortex. Label a longitudinal section and cross section of a collecting duct in the renal medulla. 🔼

Renal cortex (_____×)	Renal medulla (_____×)

Part C Assessments

Complete the following:

1. Distinguish between a renal corpuscle and a renal tubule. 🔼 _____

_____

_____

2. Number the following structures to indicate their respective positions in relation to the nephron. Assign the number 1 to the structure nearest the glomerulus. 🔼

_____   Ascending limb of nephron loop

_____   Collecting duct

_____   Descending limb of nephron loop

_____   Distal convoluted tubule

_____   Glomerular capsule

_____   Proximal convoluted tubule

_____   Papillary duct in renal papilla

3. Number the following structures (the list does not include all possible blood vessels) to indicate their respective positions in the blood pathway within the kidney. Assign the number 1 to the vessel nearest the abdominal aorta. **A**

_____   Afferent arteriole

_____   Arcuate artery

_____   Arcuate vein

_____   Cortical radiate artery

_____   Cortical radiate vein

_____   Efferent arteriole

_____   Glomerulus

_____   Peritubular capillary (or vasa recta)

_____   Renal artery

_____   Renal vein

Part D   Assessments

Sketch a cross section of a ureter and label the three layers and the lumen. Sketch a segment of a urinary bladder and label the four layers and the lumen. **5**

Ureter (_____ ✕)	Urinary bladder (_____ ✕)

Part E Assessments

Number the following structures to indicate respective positions in the pathway of urine flow in a male. Assign the number 1 to the structure nearest the papillary duct in a renal papilla. 6

_____ External urethral orifice

_____ Major calyx

_____ Membranous urethra

_____ Minor calyx

_____ Prostatic urethra

_____ Renal pelvis

_____ Spongy urethra

_____ Ureter

_____ Ureteral opening

_____ Urinary bladder

Appendix

Preparation of Solutions

Amylase solution, 0.5%

Place 0.5 g of bacterial amylase in a graduated cylinder or volumetric flask. Add distilled water to the 100 mL level. Stir until dissolved. The amylase should be free of sugar for best results; a low-maltose solution of amylase yields good results. (Store amylase powder in a freezer until mixing this solution.)

Benedict's Solution

Prepared solution is available from various suppliers.

Epsom salt solution, 0.1%

Place 0.5 g of Epsom salt in a graduated cylinder or volumetric flask. Add distilled water to the 500 mL level. Stir until dissolved.

Glucose solutions

1.0% solution. Place 1 g of glucose in a graduated cylinder or volumetric flask. Add distilled water to the 100 mL level. Stir until dissolved.

Iodine-potassium-iodide (IKI solution)

Add 20 g of potassium iodide to 1 L of distilled water, and stir until dissolved. Then add 4.0 g of iodine, and stir again until dissolved. Solution should be stored in a dark stoppered bottle.

Methylene blue

Dissolve 0.3 g of methylene blue powder in 30 mL of 95% ethyl alcohol. In a separate container, dissolve 0.01 g of potassium hydroxide in 100 mL of distilled water. Mix the two solutions. (Prepared solution is available from various suppliers.)

Monosodium glutamate (MSG) solution, 1%

Place 1 g of monosodium glutamate in a graduated cylinder or volumetric flask. Add distilled water to the 100 mL level. Stir until dissolved.

Quinine sulfate, 0.5%

Place 0.5 g of quinine sulfate in a graduated cylinder or volumetric flask. Add distilled water to the 100 mL level. Stir until dissolved.

Sodium chloride solutions

1. *0.9% solution.* Place 0.9 g of sodium chloride in a graduated cylinder or volumetric flask. Add distilled water to the 100 mL level. Stir until dissolved.
2. *1.0% solution.* Place 1.0 g of sodium chloride in a graduated cylinder or volumetric flask. Add distilled water to the 100 mL level. Stir until dissolved.
3. *3.0% solution.* Place 3.0 g of sodium chloride in a graduated cylinder or volumetric flask. Add distilled water to the 100 mL level. Stir until dissolved.
4. *5.0% solution.* Place 5.0 g of sodium chloride in a graduated cylinder or volumetric flask. Add distilled water to the 100 mL level. Stir until dissolved.

Starch solutions

1.0% solution. Add 10 g of cornstarch to 1 L of distilled water. Heat until the mixture boils. Cool the liquid, and pour it through a filter. Store the filtrate in a refrigerator.

Sucrose, 5% solution

Place 5.0 g of sucrose in a graduated cylinder or volumetric flask. Add distilled water to the 100 mL level. Stir until dissolved.

Wright's stain

Prepared solution is available from various suppliers.

Credits

NOTES

NOTES

NOTES